PHYSICAL CHEMISTRY RESEARCH FOR ENGINEERING AND APPLIED SCIENCES

VOLUME 2

Polymeric Materials and Processing

PHYSICAL CHEMISTRY RESEARCH FOR ENGINEERING AND APPLIED SCIENCES

VOLUME 2

Polymeric Materials and Processing

Edited by

**Eli M. Pearce, PhD, Bob A. Howell, PhD,
Richard A. Pethrick, PhD, DSc, and
Gennady E. Zaikov, DSc**

Apple Academic Press Inc.
3333 Mistwell Crescent
Oakville, ON L6L 0A2
Canada

Apple Academic Press Inc.
9 Spinnaker Way
Waretown, NJ 08758
USA

©2015 by Apple Academic Press, Inc.

First issued in paperback 2021

Exclusive worldwide distribution by CRC Press, a member of Taylor & Francis Group

No claim to original U.S. Government works

ISBN 13: 978-1-77463-093-8 (pbk)
ISBN 13: 978-1-77188-057-2 (hbk)

Library and Archives Canada Cataloguing in Publication

Physical chemistry research for engineering and applied sciences / edited by Eli M. Pearce, PhD, Bob A. Howell, PhD, Richard A. Pethrick, PhD, DSc, and Gennady E. Zaikov, DSc.

Includes bibliographical references and index.
Contents: Volume 2. Polymeric materials and processing.
ISBN 978-1-77188-057-2 (v. 2 : bound)
1. Chemistry, Physical and theoretical. 2. Chemistry, Technical. 3. Physical biochemistry.
I. Pearce, Eli M., author, editor II. Howell, B. A. (Bobby Avery), 1942-, author, editor
III. Pethrick, R. A. (Richard Arthur), 1942-, author, editor IV. Zaikov, G. E. (Gennadii Efremovich), 1935-, author, editor

---

QD453.3.P49 2015   541   C2015-900409-8

Library of Congress Cataloging-in-Publication Data

Physical chemistry research for engineering and applied sciences/Eli M. Pearce, PhD, Bob A. Howell, PhD, Richard A. Pethrick, PhD, DSc, and Gennady E. Zaikov, DSc.
volumes cm
Includes bibliographical references and index.
Contents: volume 1. Principles and technological implications -- volume 2. Polymeric materials and processing -- volume 3. High performance materials and methods
ISBN 978-1-77188-053-4 (alk. paper)
1. Chemistry, Physical and theoretical. 2. Chemistry, Technical. 3. Physical biochemistry. I. Pearce, Eli M. II. Howell, B. A. (Bobby Avery), 1942- III. Pethrick, R. A. (Richard Arthur), 1942- IV. Zaikov, G. E. (Gennadii Efremovich), 1935-

---

QD453.3.P49 2015   541--dc23   2015000878

Apple Academic Press also publishes its books in a variety of electronic formats. Some content that appears in print may not be available in electronic format. For information about Apple Academic Press products, visit our website at **www.appleacademicpress.com** and the CRC Press website at **www.crcpress.com**

ABOUT THE EDITORS

Eli M. Pearce, PhD

Dr. Eli M. Pearce was the President of the American Chemical Society. He served as Dean of the Faculty of Science and Art at Brooklyn Polytechnic University in New York as well as a Professor of Chemistry and Chemical Engineering. He was the Director of the Polymer Research Institute, also in Brooklyn. At present, he consults for the Polymer Research Institute. As a prolific author and researcher, he edited the *Journal of Polymer Science* (Chemistry Edition) for 25 years and was an active member of many professional organizations.

Bob A. Howell, PhD

Bob A. Howell, PhD, is a Professor in the Department of Chemistry at Central Michigan University in Mount Pleasant, Michigan. He received his PhD in physical organic chemistry from Ohio University in 1971. His research interests include flame-retardants for polymeric materials, new polymeric fuel-cell membranes, polymerization techniques, thermal methods of analysis, polymer-supported organoplatinum antitumor agents, barrier plastic packaging, bioplastics, and polymers from renewable sources.

Richard A. Pethrick, PhD, DSc

Professor R. A. Pethrick, PhD, DSc, is currently a Research Professor and Professor Emeritus in the Department of Pure and Applied Chemistry at the University of Strathclyde, Glasglow, Scotland. He was Burmah Professor in Physical Chemistry and has been a member of the staff there since 1969. He has published over 400 papers and edited and written several books. Recently, he has edited several publications concerned with the techniques for the characterization of the molar mass of polymers and also the study of their morphology. He currently holds a number of EPSRC grants and is involved with Knowledge Transfer Programmes involving three local companies involved in production of articles made out of polymeric materials. His current research involves AWE. He has acted as a consultant for BAE Systems in the area of explosives and a company involved in the production of anticorrosive coatings.

Dr. Pethrick is on the editorial boards of several polymer and adhesion journals and was on the Royal Society of Chemistry Education Board. He is a Fellow of the Royal Society of Edinburgh, the Royal Society of Chemistry, and the Institute of Materials, Metal and Mining. Previously, he chaired the 'Review of Science Provision 16-19' in Scotland and the restructuring of the HND provision in chemistry. He was also involved in the creation of the revised regulations for accreditation by the Royal Society of Chemistry of the MSc level qualifications in chemistry. For a many years, he was the Deputy Chair of the EPSRC IGDS panel and involved in a number of reviews of the courses developed and offered under this program. He has been a member of the review panel for polymer science in Denmark and Sweden and the National Science Foundation in the USA.

Gennady E. Zaikov, DSc

Gennady E. Zaikov, DSc, is the Head of the Polymer Division at the N. M. Emanuel Institute of Biochemical Physics, Russian Academy of Sciences, Moscow, Russia, and Professor at Moscow State Academy of Fine Chemical Technology, Russia, as well as Professor at Kazan National Research Technological University, Kazan, Russia.

He is also a prolific author, researcher, and lecturer. He has received several awards for his work, including the Russian Federation Scholarship for Outstanding Scientists. He has been a member of many professional organizations and on the editorial boards of many international science journals.

Physical Chemistry Research for Engineering and Applied Sciences in 3 Volumes

Physical Chemistry Research for Engineering and Applied Sciences:
Volume 1: Principles and Technological Implications
Editors: Eli M. Pearce, PhD, Bob A. Howell, PhD,
Richard A. Pethrick, PhD, DSc, and Gennady E. Zaikov, DSc

Physical Chemistry Research for Engineering and Applied Sciences:
Volume 2: Polymeric Materials and Processing
Editors: Eli M. Pearce, PhD, Bob A. Howell, PhD,
Richard A. Pethrick, DSc, PhD, and Gennady E. Zaikov, DSc

Physical Chemistry Research for Engineering and Applied Sciences:
Volume 3: High Performance Materials and Methods
Editors: Eli M. Pearce, PhD, Bob A. Howell, PhD,
Richard A. Pethrick, DSc, PhD, and Gennady E. Zaikov, DSc

CONTENTS

LIST OF CONTRIBUTORS

Yu. O. Andriasyan
N. M. Emanuel Institute of Biochemical Physics, Russian Academy of Sciences, 4 Kosygin str., 119334 Moscow, Russia

J. N. Aneli
Laboratory for Polymer materials, Institute of Machine Mechanics, Mindeli Str.10, Tbilisi 0186, Republic Georgia, E-mail: jimaneli@yahoo.com

A. Antonov
N. M. Emanuel Institute of Biochemical Physics, Russian Academy of Sciences, Kosygin Str. 4, 119334 Moscow, Russia

Marina Bazunova
Bashkir State University 32 Zaki Validi Street, 450076 Ufa, Republic of Bashkortostan, Russia

Dariusz M. Bieliński
Institute for Engineering of Polymer Materials & Dyes, Department of Elastomers and Rubber Technology, Harcerska 30, 05-820 Piastów, Poland, tel. +4842 6313214, fax +4842 6362543, E-mail: dbielin@p.lodz.pl

R. Y. Deberdeev
Kazan National Research Technological University, 65 Karl Marx str., Kazan 420015, Tatarstan, Russia

Jacek Grams
Institute of General and Ecological Chemistry, Technical University of Łódź, Stefanowskiego 12/16, 90-924 Łódź, Poland

D. O. Gusev
Volgograd State Technical University, 400005, 28 Lenin ave., Volgograd, Russia

R. Jozwik
Military Institute of Chemistry and Radiometry, Al. gen. A. Chrusciela "Montera" 105, 00-910 Warsaw, Poland

Zinaida S. Khasbulatova
Chechen State Pedagogical Institue, 33 Kievskaya str., Groznyi 364037, Chechnya, Russia

Ivan Krupenya
Bashkir State University 32 Zaki Validi Street, 450076 Ufa, Republic of Bashkortostan, Russia

Elena Kulish
Bashkir State University 32 Zaki Validi Street, 450076 Ufa, Republic of Bashkortostan, Russia

E. M. Kuvardina
The South-West state university, 305040, Kursk, to St. 50-years of October, 94

N. V. Kuvardin
The South-West state university, 305040, Kursk, to St. 50-years of October, 94

N. G. Lebedev
Volgograd State University, Volgograd, Russia

D. V. Medvedev
Elastomer Limited Liability Company, 400005, 75 Chuikova st., Volgograd, Russia

G. V. Medvedev
Volgograd State Technical University, 400005, 28 Lenin ave., Volgograd, Russia

I. A. Mikhaylov
N. M. Emanuel Institute of Biochemical Physics, Russian Academy of Sciences, 4 Kosygin str., 119334 Moscow, Russia

Vadim Z. Mingaleev
Institute of Organic Chemistry, Ufa Scientific Center of Russian Academy of Sciences, pr. Oktyabrya 71, Ufa, Bashkortostan, 450054, Russia

P. Moldenaers
K. U. Leuven, Department of Chemical Engineering, W. de Croylaan 46, B-3001 Leuven, Belgium

T. M. Natriashvili
Laboratory for Polymer materials, Institute of Machine Mechanics, Mindeli Str.10,Tbilisi 0186 , Republic Georgia,

F. F. Niyazy
The South-West state university, 305040, Kursk, to St. 50-years of October, 94, E-mail: farukhni-yazi@yandex.com

I. A. Novakov
Volgograd State Technical University, 400005, 28 Lenin ave., Volgograd, Russia

A. A. Popov
N. M. Emanuel Institute of Biochemical Physics, Russian Academy of Sciences, 4 Kosygin str., 119334 Moscow, Russia

P. Van Puyvelde
K. U. Leuven, Department of Chemical Engineering, W. de Croylaan 46, B-3001 Leuven, Belgium

Mariusz Siciński
Institute of Polymer and Dye Technology, Technical University of Łódź, Stefanowskiego 12/16, 90-924 Łódź, Poland

N. V. Sidorenko
Volgograd State Technical University, 400005, 28 Lenin ave., Volgograd, Russia

S. A. Sudorgin
Volgograd State Technical University, Volgograd, Russia

M. A. Vaniev
Volgograd State Technical University, 400005, 28 Lenin ave., Volgograd, Russia, E-mail: vaniev@vstu.ru

Michał Wiatrowski
Department of Molecular Physics, Technical University of Łódź, Stefanowskiego 12/16, 90-924 Łódź, Poland

Gennady E. Zaikov
N. M. Emanuel Institute of Biochemical Physics, Russian Academy of Sciences, 4 Kosygin str., 119334 Moscow, Russia, E-mail: Chembio@sky.chph.ras.ru

Elena M. Zakharova
Institute of Organic Chemistry, Ufa Scientific Center of Russian Academy of Sciences, pr. Oktyabrya 71, Ufa, Bashkortostan, 450054, Russia

Vadim P. Zakharova
Bashkir State University, Zaki Validi str. 32, Ufa, 450076 Bashkortostan, Russia

Iriva D. Zakirovaa
Bashkir State University, Zaki Validi str. 32, Ufa, 450076 Bashkortostan, Russia

LIST OF ABBREVIATIONS

AFM	Atomic Force Microscopy
AP	Aromatic Polyesters
CNT	Carbon Nanotubes
CTZ	Polysaccharide Chitosan
DCV- GCMD	Dual-Volume GCMD
DM	Dibenzothiazole Disulphide
DP	Diamond Pore
DSC	Differential Scanning Calorimetry
EB	Electron Beam
ER	Epoxy Resin
F	Flexible
FG	Fiber Glass
GCMD	Grand Canonical Molecular Dynamics
HR	Heat Radiation
HTS	High Temperature Shearing
IUPAC	International Union of Pure and Applied Chemistry
LDPE	Low Density Polyethylene
LQPS	Liquid-Crystal Polyesters
MC	Monte Carlo
MD	Molecular Dynamics
MF	Microfiltration
MSD	Mean-Square Displacement
MWCO	Molecular Weight Cut-Off
MWNT	Multi-Walled Carbon Nanotube
NBR	Acrylonitrile-Butadiene Rubber
NCN	National Science Centre Poland
NF	Nanofiltration
NR	Natural Rubber
OIT	Oxidation Induction Time
OOT	Oxidation Onset Temperature
PEG	Polyethylene Glycols
PFR	Phenoloformaldehyde Resin
PMS	Polymethyl-Silsesquioxane
PP	Polypropylene
PUE	Polyurethane Elastomers

R	Rigid
RESPA	Reference System Propagator Algorithm
RHR	Rate of Heat Release
RO	Reverse Osmosis
ROA	Rheometrics Optical Analyzer
SALS	Small Angle Light Scattering
SBR	Styrene-Butadiene Rubber
SE	Secondary Electron Signal
SEM	Scanning Electron Microscope
SFE	Surface Free Energy
SP	Straight Path
SWNT	Single-Walled Carbon Nanotube
TPES	Thermoplastic Elastomers
TS	Tensile Strength
UF	Ultrafiltration
US	Ultrasound
VACF	Velocity Autocorrelation Function
ZP	Zigzag Path

LIST OF SYMBOLS

α_0	amplitude of the initial disturbance
α_B	amplitude of the instability
$\bar{u}$	average molecular speed
$A_{m's}$	coefficients of the Fourier expansion
P_S^*	constant
$c_{j\sigma}^+$	creation operators of electrons
t_Δ	electron hopping integral
$\varepsilon_{l\sigma}$	energy of the electron by the impurity
$c_{j\sigma}$	Fermi annihilation
$f_s(p,r)$	Fermi distribution function
f_A	fugacity
V_{lj}	matrix element of hybridization
ξ_i	random numbers generated for each trial
C_h^*	saturation constant
D_A^*	self-diffusion coefficient
Γ	surface tension
η_m	viscosity of the matrix material
A	proportionality constant
a_1 and a_2	unit vectors
aq	energy barrier
B	hole affinity constant
$c(x)$	concentration
c_A	concentration of diffusant A
D	pore diameter
e_1, e_2 and e_3	coordinates in the current configuration
eV	electron energy
F	applied external force
G_1, G_2	material coordinates of a point in the initial configuration
J	molecular flux
K	temperature dependent Henry's law coefficient
K_n	Knudsen number
K_0	proportionality constant
K_P	henry's law constant
L	membrane thickness
M	molecular mass

m_{sample}	sample weight
M_{η}	molecular weight
N	number of carbon atoms in the lattice
n	quantization number
Na	the number of molecules
N_{imp}	number of adsorbed hydrogen atoms
P	permeability
p_x	parallel component of the graphene sheet
Q	heat of adsorption
R	radius of the modeled SWCNT
R_0	initial radius of the undisturbed fibril
r_p	pore radius
S	solubility coefficient
T	thickness of the adsorbate film
T_0	constants depending on some quantum mechanical values
T_{melt}	melting point of polyamide
U	constant of the Coulomb repulsion
V	center-of-mass velocity component
V	hybridization potential
V'', V'	volumes of polysaccharide
V_L	molecular volume of the condensate
X	position across the membrane
Γ	interfacial tension
Δp	pressure drop across the membrane
ΔS	entropy change
ΔH	fusion heat
Θ	time-lag
Λ	mean free path of molecules
$\rho(x)$	an arbitrary probability distribution function
ρ'', ρ'	density of polysaccharide
T	pore tortuosity
Ω	known function of the viscosity ratio

PREFACE

Polymers play a significant part in human's existence. They have a role in every aspect of modern life, such as health care, food, information technology, transportation, energy industries, etc. The speed of developments within the polymer sector is phenomenal and, at the same time, crucial to meet demands of today's and future life. Specific applications for polymers range from adhesives, coatings, painting, foams, and packaging to structural materials, composites, textiles, electronic and optical devices, biomaterials, and many others with uses in industry and daily life. Polymers are the basis of natural and synthetic materials. They are macromolecules and in nature are the raw material for proteins and nucleic acids, which are essential for human bodies.

Cellulose, wool, natural rubber and synthetic rubber, and plastics are well-known examples of natural and synthetic types. Natural and synthetic polymers play a massive role in everyday life, and a life without polymers really does not exist.

Previously, it was believed that polymers could only be prepared through addition polymerization. The mechanism of the addition reaction was also unknown, and hence there was no sound basis of proposing a structure for polymers. Studies by researchers resulted in theorizing the condensation polymerization. This mechanism became well understood, and the structure of the resultant polyester could be specified with greater confidence.

In 1941–42, the world witnessed the infancy of polyethylene terephthalate, better known as the polyester. A decade later for the first time polyester/cotton blends were introduced. In those days Terylene and Dacron (commercial names for polyester fibers) were miracle fibers but were still overshadowed by nylon. Not many would have predicted that decades later, polyester would have become the world's inexpensive, general purpose fiber as well as becoming a premium fiber for special functions in engineering textiles, fashion, and many other technical end uses. From the time nylon and polyester were first used, there have been amazing technological advances which have made them so cheap to manufacture and widely available.

These developments have made polymers such as polyesters contribute enormously to today's modern life. One of the most important applications is the furnishing sector (home, office, cars, aviation industry, etc.) which benefits hugely from the advances in technology. There are a number of requirements for a fabric to function for its chosen end use, for example, resistance to pilling and abrasion, as well as dimensional stability. Polyester is now an important part of upholstery fabrics. The shortcomings attributed to fiber in its early days have mostly been overcome. Now it plays a significant part in improving the lifespan of a fabric as well as its dimensional stability which is due to its heat-setting properties.

About a half century has passed since synthetic leather, a composite material completely different from conventional ones, came to the market. Synthetic leather was originally developed for end uses such as the upper of shoes. Gradually other uses like clothing steadily increased the production of synthetic leather and suede. Synthetic leathers and suede have a continuous ultrafine porous structure comprising a three-dimensional entangled nonwoven fabric and an elastic material principally made of polyurethane. Polymeric materials consisting of the synthetic leathers are polyamide and polyethylene terephthalate for the fiber and polyurethanes with various soft segments, such as aliphatic polyesters, polyethers and polycarbonates, for the matrix.

New applications are being developed for polymers at a very fast rate all over the world at various research centers. Examples of these include electro active polymers, nano products, robotics, etc. Electro active polymers are special types of materials, that can be used, for example, as artificial muscles and facial parts of robots or even in nano robots. These polymers change their shape when activated by electricity or even by chemicals. They are lightweight but can bear a large force, which is very useful when being utilized for artificial muscles. Electro active polymers together with nanotubes can produce very strong actuators. Currently research work is carried out to combine various types of electro active polymers with carbon nanotubes to make the optimal actuator. Carbon nanotubes are very strong, and elastic, and they conduct electricity. When they are used as an actuator in combination with an electro active polymer the contractions of the artificial muscle can be controlled by electricity. Already work is under way to use electro active polymers in space. Various space agencies are investigating the possibility of using these polymers in space. This technology has a lot to offer for the future, and with the ever-increasing work on nanotechnology, electro active materials will play a very important part in modern life.

This book presents some fascinating phenomena associated with the remarkable features of high performance polymers and also provides an update on applications of modern polymers. It offers new research on structure–property relationships, synthesis and purification, and potential applications of high performance polymers. The collection of topics in this book reflects the diversity of recent advances in modern polymers with a broad perspective that will be useful for scientists as well as for graduate students and engineers. The book helps to fill the gap between theory and practice. It explains the major concepts of new advances in high performance polymers and their applications in a friendly, easy-to-understand manner. Topics include high performance polymers and computer science integration in biochemical, green polymers, molecular nanotechnology and industrial chemistry. The book opens with a presentation of classical models, moving on to increasingly more complex quantum mechanical and dynamical theories. Coverage and examples are drawn from modern polymers.

CHAPTER 1

INVESTIGATION ON THE INFLUENCE OF A STRONG ELECTRIC FIELD ON THE ELECTRICAL, TRANSPORT AND DIFFUSION PROPERTIES OF CARBON NANOSTRUCTURES

S. A. SUDORGIN and N. G. LEBEDEV

CONTENTS

ABSTRACT

Examines the influence of defects on the electrical properties of carbon nano-structures in an external electric field. Defects are the hydrogen atoms, which adsorbed on the surface of carbon nanostructures. Carbon nanostructures are considered the single-walled zigzag carbon nanotubes. Atomic adsorption model of hydrogen on the surface of single-walled zigzag carbon nanotubes based on the single-impurity Anderson periodic model. Theoretical calculation of the electron diffusion coefficient and the conductivity of zigzag carbon nanotubes alloy hydrogen atoms carried out in the relaxation time approximation. Revealed a decrease in the electrical conductivity and the electron diffusion coefficient with increasing concentration of adsorbed hydrogen atoms. The nonlinearity of the electrical conductivity and the diffusion coefficient of the amplitude of a constant strong electric field at the constant concentration of hydrogen adatoms shown at the figures.

1.1   INTRODUCTION

Despite the already long history of the discovery of carbon nanotubes (CNT) [1] the interest in the problem of obtaining carbon nanostructures with de-sired characteristics unabated, constantly improving their synthesis. Unique physical and chemical properties of CNTs can be applied in various fields of modern technology, electronics, materials science, chemistry and medicine [2]. One of the most important from the point of view of practical applications are the transport properties of CNTs.

Under normal conditions, any solid surfaces coated with films of atoms or molecules adsorbed from the environment, or left on the surface in the diffusion process [3]. The most of elements adsorption on metals forms a chemical bond. The high reactivity of the surface of carbon nanotubes makes them an exception. Therefore, current interest is the study of the influence of the adsorption of atoms and various chemical elements and molecules on the electrical properties of carbon nanostructures.

In the theory of adsorption, in addition to the methods of quantum chemistry, widely used the method of model Hamiltonians [3]. In the study of the adsorption of atoms and molecules on metals used primarily molecular orbital approach-self-consistent field, as this takes into account the delocalization of electrons in the metal. Under this approach, the most commonly used model Hamiltonian Anderson [4, 5], originally proposed for the description of the electronic states of impurity atoms in the metal alloys. The model has been successfully applied to study the adsorption of atoms on the surface of metals and semiconductors [6], the adsorption of hydrogen on the surface of graphene [7] and carbon nanotubes [8, 9].

In this paper we consider the influence of the adsorption of atomic hydrogen on the conducting and diffusion properties of single-walled "zigzag" CNTs. Interaction of hydrogen atoms adsorbed to the surface of carbon nanotubes is described in terms of the periodic Anderson model. Since the geometry of the CNT determines their conductive properties, then to describe the adsorption on the surface of CNTs using this model is justified. Transport coefficients (conductivity and diffusion coefficient) CNT electron calculated by solving the Boltzmann equation [10] in the relaxation time approximation.

This technique was successfully applied by authors to calculate the ideal transport characteristics of carbon nanotubes [11], graphene bilayer graphene [12] and graphene nanoribbons [13].

1.2 MODEL AND BASIC RELATIONS

However, with the discovery of new forms of carbon model can be successfully applied to study of the statistical properties of CNTs and graphene. Carbon atom in the nanotube forms three chemical connection σ-type. Lodging with nearest neighbor atoms with three sp^2 hybridization of atomic orbitals. The fourth p-orbital involved in chemical bonding π-type which creates π-shell nanotube describing state of itinerant electrons, that define the basic properties of CNTs and graphene. This allows us to consider the state of π-electron system in the framework of the Anderson model. The model takes into account the kinetic energy of electrons and their Coulomb interaction at one site and neglected energy inner-shell electrons of atoms and electrons involved in the formation of chemical bonds σ-type, as well as the vibrational energy of the atoms of the crystal lattice.

In general, the periodic Anderson model [5] considers two groups of electrons: itinerant s-electrons and localized d-electrons. Itinerant particles are considered free and localized-interact by Coulomb repulsion on a single node. With the discovery of new forms of carbon model can be successfully applied to study the statistical properties of carbon structures are the CNT and the graphene. Carbon atom in the graphene layer has 3 forms chemical bonds σ-type with its immediate neighbors. The fourth orbital p-type forms a chemical bond π-type, describing the state of itinerant electrons. States localized electrons created by the valence orbitals (in this case, the p-type) impurity atoms. This allows us to consider the state of π-electrons in the framework of the Anderson model. The model takes into account the kinetic energy of the electrons in the crystal and impurity electrons interacting through a potential hybridization, and neglects the energy of the electrons of the inner shells of atoms and electrons involved in the formation of chemical bonds σ-type, as well as the vibrational energy of the atoms of the crystal lattice [5].

In the periodic Anderson model state of the electrons of the crystal containing impurities in the π-electron approximation and the nearest neighbor approximation is described by the effective Hamiltonian, having the following standard form [5]:

$$H = \sum_{j,\Delta,\sigma} t_\Delta \left(c_{j\sigma}^+ c_{j+\Delta\sigma} + c_{j+\Delta\sigma}^+ c_{j\sigma} \right) + \sum_{l,\sigma} \varepsilon_{l\sigma} n_{l\sigma}^d + \sum_l U n_{l\uparrow}^d n_{l\downarrow}^d +$$
$$+ \sum_{l,j,\sigma} \left(V_{lj} c_{j\sigma}^+ d_{l\sigma} + V_{lj}^* d_{l\sigma}^+ c_{j\sigma} \right) \qquad , \qquad (1)$$

where t_Δ is the electron hopping integral between the neighboring lattice sites of the crystal; U is the constant of the Coulomb repulsion of the impurity; $c_{j\sigma}$ and $c_{j\sigma}^+$ are the Fermi annihilation and creation operators of electrons in the crystal node j with spin σ; $d_{j\sigma}$ and $d_{j\sigma}^+$ are the Fermi annihilation and creation operators of electrons on the impurities l with spin σ; $n_{l\sigma}^d$ is the operator of the number of electrons on impurities l with spin σ; $\varepsilon_{l\sigma}$ is the energy of the electron by the impurity l with spin σ; V_{lj} is the matrix element of hybridization of impurity electron l and atom j of the crystal.

After the transition to k-space by varying the crystal by Fourier transformation of creation and annihilation of electrons and crystal use the Green function method, the band structure of single-walled CNTs with impurities adsorbed hydrogen atoms takes the form [8, 9]:

$$E(\mathbf{k}) = \frac{1}{2} \left[\varepsilon_k + \varepsilon_{l\sigma} \pm \left(\left(\varepsilon_k - \varepsilon_{l\sigma} \right)^2 + 4 \frac{N_{imp}}{N} |V|^2 \right)^{\frac{1}{2}} \right], \qquad (2)$$

where N – number of carbon atoms in the lattice, determines the size of the crystal, N_{imp} – the number of adsorbed hydrogen atoms, V – hybridization potential, $\varepsilon_{l\sigma} = -5.72$ eV – electron energy impurities-the band structure of an ideal single-walled nanotubes, for tubes, for example, zigzag type dispersion relation is defined as follows [1]:

$$E(\mathbf{p}) = \pm\gamma\sqrt{1 + 4\cos\left(ap_x\right)\cos\left(\pi s/n\right) + 4\cos^2\left(\pi s/n\right)} \qquad (3)$$

where $a = 3d/2\hbar$, $d = 0.142$ nm is the distance between adjacent carbon atoms in graphene, $\mathbf{p} = (p_x, s)$ is the quasi momentum of the electrons in graphene, p_x is the parallel component of the graphene sheet of the quasi momentum and $s = 1, 2, \ldots, n$ are the quantization numbers of the momentum components depending on the width of the graphene ribbon. Different signs are related to the conductivity band and to the valence band accordingly.

Used in the calculation of the Hamiltonian parameters: the value of the hopping integral $t_0 = 2.7$ eV, hybridization potential $V = -1.43$ eV estimated from quantum chemical calculations of the electronic structure of CNTs within the semiempirical MNDO [14]. Electron energy impurity $\varepsilon_{1\sigma} = -5.72$ eV was assessed using the method described in Refs. [6, 7].

Consider the effect of the adsorption of atomic hydrogen on the response of single-walled "zigzag" CNTs to an external electric field applied along the x axis is directed along the axis of the CNT (Fig. 1.1).

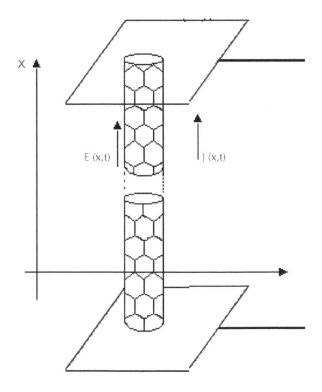

FIGURE 1.1 Geometry configuration. CNT type zigzag is in an external electric field. Field strength vector E is directed along the axis of the CNT.

Method of calculating the transport coefficients of electrons in carbon nanotubes described in detail in Refs. [11–13]. Evolution of the electronic system was simulated in the semi classical approximation of the relaxation time. Electron distribution function in the state with momentum $p = (p_x, s)$ is of the t-approximation using Boltzmann equation [10]:

$$\frac{\partial f_s(\mathbf{p},\mathbf{r})}{\partial t} + \mathbf{F}\frac{\partial f_s(\mathbf{p},\mathbf{r})}{\partial \mathbf{p}} = \frac{f_s(\mathbf{p},\mathbf{r}) - f_{0s}(\mathbf{p},\mathbf{r})}{\tau}, \qquad (4)$$

where $f_s(\mathbf{p},\mathbf{r})$-the Fermi distribution function $\mathbf{F} = e\mathbf{E}$ -acting on the particle constant electrostatic force.

To determine the dependence of the diffusion and conductive character-istics of CNTs on the external electric field using the procedure outlined in Ref. [15]. The longitudinal component of the current density $j = j_x$ has the following form:

$$j(x) = \sigma(\mathbf{E})\mathbf{E} + D(\mathbf{E})\frac{\nabla_x n}{n} \qquad (5)$$

For the case of a homogeneous temperature distribution $T(r) =$ constant in the linear approximation in magnitude [11], expressions for the transport coef-ficients of single-walled nanotubes: conductivity and diffusivity of electrons. Electrical conductivity of CNT type zigzag given following expression [11]:

$$\sigma(E) = \sum_s \int_{-\pi}^{\pi} \sum_m A_{ms} m f_{0s}(p_x, x)\frac{E}{E^2 m^2 + 1}\left[\sin(mp_x) + Em\cos(mp_x)\right] dp_x \qquad (6)$$

Expression for the diffusion coefficient of electrons in CNT type zigzag has the form [11]:

$$D(E) = \sum_s \int_{-\pi}^{\pi} f_{0s}(p_x, x)\sum_m A_{ms} m \sum_{m'} A_{m's} m'\left\{\frac{[E^2(m^2 + m'^2) + 1][EmR + M]}{K} + \right.$$

$$\left. + \frac{[E^3(m'^3 - 2m^2 m') + Em']T}{K}\right\} dp_x + \sum_s \int_{-\pi}^{\pi} f_{0s}(p_x, x)\sum_m A_{ms} m \sum_{m'} A_{m's} m'\frac{F}{P} dp_x, \qquad (7)$$

where the following notation:

$$K = [E^4(m^4 + m'^4 - 2m^2 m'^2) + 2E^2(m^2 + m'^2) + 1][E^2 m^2 + 1],$$

$$P = [E^2 m^2 + 1]^2 [E^2 m'^2 + 1]$$,

$$R = \cos(mp_x)\sin(m'p_x) + \cos(mp_x)\cos(m'p_x) - \sin(mp_x)\sin(m'p_x),$$

$$M = \sin(mp_x)\sin(m'p_x) + \sin(mp_x)\cos(m'p_x) + \cos(mp_x)\sin(m'p_x),$$

$$T = [\cos(mp_x)\cos(m'p_x) - Em\sin(mp_x)\cos(m'p_x)],$$

$$F = [\sin(m'p_x) + Em\cos(m'p_x)][\sin(mp_x) + 2Em\cos(mp_x) - E^2 m^2 \sin(mp_x)],$$

$A_{ms}, A_{m's}$ are the coefficients of the Fourier expansion of the dispersion relation of electrons in CNT, m and m 'order Fourier series. For the convenience of visualization and qualitative analysis performed procedure and select the following dimensionless relative unit of measurement of the electric field $E_0 = 4.7 \times 10^6$ V/m.

1.3   RESULTS AND DISCUSSION

To investigate the influence of an external constant electric field on the transport properties of single-walled CNT type zigzag with adsorbed hydrogen atoms selected the following system parameters: temperature T ≈ 300 K, the relaxation time is $\tau \approx 10^{-12}$ s in accordance with the data [16]. For numerical analysis considered type semiconducting CNT (10,0).

It should be noted that a wide range of external field behavior of the specific conductivity σ(E) for nanotubes with hydrogen adatoms has the same qualitative nonlinear dependence as for the ideal case of nanoparticles, which was discussed in detail in Ref. [11]. In general, the dependence of conductivity on the electric field has a characteristic for semiconductors form tends to saturate and decreases monotonically with increasing intensity. This phenomenon is associated with an increase in electrons fill all possible states of the conduction band. Behavior of electrical conductivity under the influence of an external electric field is typical for semiconductor structures with periodic and limited dispersion law [17].

Figure 1.2 shows the dependence of conductivity σ(E) on the intensity of the external electric field E for ideal CNT (10,0) and CNT (10,0) with adsorbed hydrogen at relatively low fields. The graphs show that the addition of single adsorbed atom (adatom) hydrogen reduces the conductivity by a small amount (about 2×10⁻³ S/m). Lowering the conductivity of the hydrogen atom in the adsorption takes place due to the fact that one of the localized electron crystallite forms a chemical bond with the impurity atom and no longer participates in the charge transport by CNT.

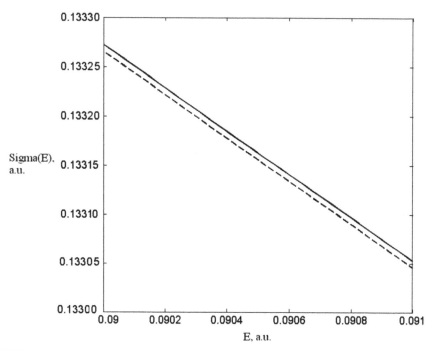

FIGURE 1.2 Dependence of the conductivity $\sigma(E)$ on the magnitude of tension external electric field E: for ideal CNT (10,0)-solid line and the CNT (10,0) with hydrogen adatom-dashed line. x-axis is a dimensionless quantity of the external electric field E (unit corresponds to 4.7×10^6 V/m), the y-axis is dimensionless conductivity $\sigma(E)$ (unit corresponds to 1.9×10^3 S/m).

Also analyzed the dependence of the conductivity $\sigma(E)$ on the intensity of the external electric E for CNT (10,0) type, containing different concentrations of hydrogen adatoms (Fig. 1.3). The increasing of the number of adsorbed atoms reduces the conductivity of zigzag CNT proportional to the number of localized adsorption bonds formed. When you add one hydrogen adatom conductivity of CNT type (10,0) is reduced by 0.06%, adding 100 adatoms by 0.55%, adding 300 adatoms by 1.66%, adding 500 adatoms by 2.62%.

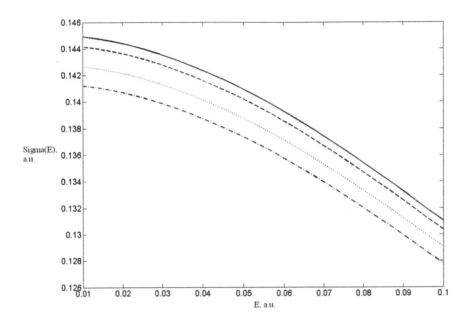

FIGURE 1.3 Dependence of the conductivity σ(E) on the magnitude of tension E external electric impurity for CNT (10,0) one hydrogen adatom-solid line; 100 adatoms-dashed line; 300 adatoms-dotted line; 500 adatoms-dash-dot line. x-axis is a dimensionless quantity of the external electric field E (unit corresponds to 4.7×10⁶ V/m), the y-axis is dimensionless conductivity σ(E) (unit corresponds to 1.9×10³ S/m).

Figure 1.4 shows that this behavior is typical for semiconductor conductivity of CNTs with different diameters. With the increasing diameter of the nanotubes have high electrical conductivity, since they contain a larger amount of electrons, which may participate in the transfer of electrical charge. The graphs in Fig. 1.4 shows for the (5,0), (10,0) and (20,0) CNT with the addition of 100 hydrogen adatoms.

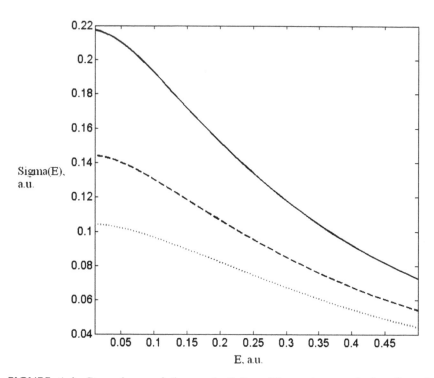

FIGURE 1.4 Dependence of the conductivity $\sigma(E)$ on the magnitude of tension external electric E for different types of CNTs with the addition of hydrogen adatoms 100 (20,0)-solid line, (10,0)-dashed line; (5,0)-the dotted line. x-axis is a dimensionless quantity of the external electric field E (unit corresponds to 4.7×10^6 V/m), the y-axis is dimensionless conductivity $\sigma(E)$ (unit corresponds to 1.9×10^3 S/m).

The electron diffusion coefficient $D(E)$ from the electric field in the single-walled zigzag CNT with adsorbed hydrogen atoms has a pronounced nonlinear character (Fig. 1.5). Increase of the field leads to an increase in first rate, and then to his descending to a stationary value. This phenomenon is observed for all systems with intermittent and limited electron dispersion law [17]. Electron diffusion coefficient can be considered constant in the order field amplitudes $E \approx 5 \times 10^6$ V/m. The maximum value of the diffusion coefficient for semiconductor CNTs observed at field strengths of the order of $E \approx 4.8 \times 10^5$ V/m.

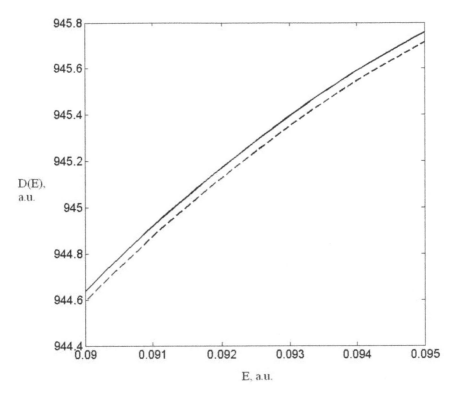

FIGURE 1.5   Dependence of the electron diffusion coefficient $D(E)$ on the intensity of the external electric field E: for CNT (10,0) ideal-solid line and hydrogen adatom-dashed line. x-axis is a dimensionless quantity of the external electric field E (unit corresponds to 4.7×10^6 V/m), the y-axis is a dimensionless diffusion coefficient $D(E)$ (unit corresponds to 3.5×10^2 A/m).

When adding the adsorbed hydrogen atoms the electron diffusion coefficient, as well as the conductivity is reduced by 0.05% (Fig. 1.5). This behavior of the diffusion coefficient in an external electric field is observed for different concentrations of hydrogen adatoms (Fig. 1.6) and semiconductor CNTs with different diameters by adding 100 adatoms (Fig. 1.7).

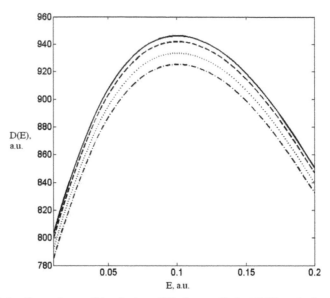

FIGURE 1.6   Dependence of the electron diffusion coefficient $D(E)$ on the intensity of the external electric E for impurity CNT (10,0) one hydrogen adatom-solid line; 100 adatoms-dashed line; 300 adatoms-dotted line; 500 adatoms-dash-dot line. x-axis is a dimensionless quantity of the external electric field E (unit corresponds to 4.7×10^6 V/m), the y-axis is a dimensionless diffusion coefficient $D(E)$ (unit corresponds to 3.5×10^2 A/m).

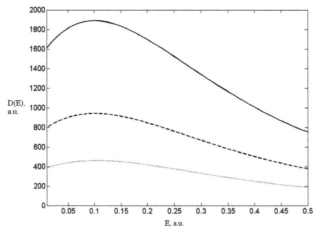

FIGURE 1.7   Dependence of the electron diffusion coefficient $D(E)$ on the intensity of the external electric E for different types of CNTs with the addition of hydrogen adatoms 100 (20,0)-solid line, (10,0)-dashed line; (5,0)-the dotted line. x-axis is a dimensionless quantity of the external electric field E (unit corresponds to 4.7×10^6 V/m), the y-axis is a dimensionless diffusion coefficient $D(E)$ (unit corresponds to 3.5×10^2 A/m).

The presented results can be used for the preparation of carbon nanotubes with desired transport characteristics and to develop the microelectronic devices, which based on carbon nanoparticles.

1.4 CONCLUSION

The main results are formulated in the conclusion.
1. The method for theoretical calculation of the semiconducting zigzag CNT transport properties with adsorbed hydrogen atoms developed. Analytical expressions for the conductivity and the electron diffusion coefficient in zigzag CNT with hydrogen adatoms in the presence of an electric field.
2. Numerical calculations showed nonlinear dependence of the transport coefficients on the electric field. For strong fields coefficients tend to saturate.
3. Atomic hydrogen adsorption of the semiconducting zigzag CNT reduces their conductivity by several percent. The electron diffusion coefficient also decreases with increasing concentration of adsorbed hydrogen atoms, and a decrease of the diffusion coefficient is more pronounced than the decrease of electrical conductivity for each of the above types of semiconducting CNTs at a larger number of adatoms.
4. Transport properties of nanotubes with adatoms increases with the diameter. A physical explanation for the observed effect.

ACKNOWLEDGMENT

This work was supported by the Russian Foundation for Basic Research (Grant № 13–03–97108, Grant № 14–02–31801), and the Volgograd State University grant (Project № 82–2013-a/VolGU).

KEYWORDS

- Adsorption
- Carbon nanostructures
- Conductivity
- Diffusion coefficient
- "zigzag" nanotubes

REFERENCES

1. Diachkov, P. N. (2010). Electronic Properties and Applications of Nanotubes. M. Bin, the Laboratory of Knowledge, 488 p.
2. Roco, M. C., Williams, R. S. & Alivisatos, A. P. (2002). Nanotechnology in the Next Decade. Forecast the Direction of Research. Springer-Verlag, 292 p.
3. Bolshov, L. A., Napartovich, A. P., Naumovets, A. G. & Fedorus, A. G. (1977). UFN. T. *122(1)*. 125.
4. Anderson, P. W. (1961). *Phys. Rev.*, *124*, 41.
5. Izyumov, A., Chashchin, I. I. & Alekseev, D. S. (2006). The Theory of Strongly Correlated Systems. Generating Functional Method. Moscow-Izhevsk: NITs "*Regular and Chaotic Dynamics*," 384.
6. Davydov, S. U. & Troshin, S. V. (2007*). Phys. T. 49(8)*. 1508.
7. Davydov, S. Y. & Sabirov, G. I. (2010). *Letters* ZHTF. T. *36(24)*. 77.
8. Pak, A. V. & Lebedev, N. G. (2012). *Chemical Physics*, *31(3)*, 82–87.
9. Pak, A. V. & Lebedev, N. G. (2013). *ZhFKh. T. 87*, 994.
10. Landau, L. D. & Lifshitz, E. M. (1979). Physical Kinetics. *M. Sci. Lit.*, 528 c.
11. Belonenko, M. B., Lebedev, N. G. & Sudorgin, S. A. (2011). *Phys. T.* 53. S. 1841.
12. Belonenko, M. B., Lebedev, N. G. & Sudorgin, S. A. (2012). *Technical Physics*, *82(7)*, 129–133.
13. Sudorgin, S. A., Belonenko, M. B. & Lebedev, N. G. (2013). *Physica Scripta*, *87(1)*, 15–602 (1–4).
14. Stepanov, N. F. (2001). Quantum Mechanics and Quantum Chemistry. Springer-Verlag, 519 sec.
15. Buligin, A. S., Shmeliov, G. M. & Maglevannaya, I. I. (1999). *Phys. T. 41*, 1314.
16. Maksimenko, S. A. & Slepyan, G. Ya. (2004). Nanoelectromagnetics of Low-Dimensional Structure. In Handbook of Nanotechnology. Nanometer Structure: Theory, Modeling, and Simulation. Bellingham: SPIE Press, 145 p.
17. Dykman, I. M. & Tomchuk, P. M. (1981). Transport Phenomena and Fluctuations in Semiconductors. Kiev. *Sciences*. Dumka, 320 c.

CHAPTER 2

A STUDY THERMAL STABILITY OF POLYURETHANE ELASTOMERS

I. A. NOVAKOV, M. A. VANIEV, D. V. MEDVEDEV,
N. V. SIDORENKO, G. V. MEDVEDEV, and D. O. GUSEV

CONTENTS

ABSTRACT

Thermal stability of polyurethane elastomers based on a product of the anionic copolymerization of butadiene and isoprene in the ratio of 80:20 and isoprene was first studied by DSC. The preferred conditions (temperature of the isothermal segment and oxygen consumption) were revealed to determine the oxidation induction time of this type of materials. The effect of Irganox 1010, Evernox 10, Songnox 1010 and 1010 Chinox stabilizers on the oxidation induction time has been studied.

It was found that the products of the same chemical structure, depending on commercial brands, may display different antioxidant activity in OIT tests. It was pointed out that this factor must be taken into account in developing polyurethane composition formulations.

2.1   INTRODUCTION

Polyurethane elastomers (PUE) are of great practical importance in various fields [1]. In particular, in developing PUE of molding compositions for sports and roofing the liquid rubbers (oligomers) of diene nature with a molecular weight of 2000–4000 are widely used as a polyol component. Usually these are homopolymers of butadiene and isoprene, the products of copolymerization of butadiene with isoprene or butadiene with piperylene and isocyanate prepolymers based on these oligomers.

After curing, the materials exhibit good physical-mechanical, dynamic and relaxation properties, high hydrolytic stability [2, 3]. However, the disadvantage of these PUEs is their low resistance to thermal-oxidation aging, due to the presence of double bonds in the oligomer molecules. Under the effect of weather conditions, irreversible changes leading to partial or complete loss of the fundamental properties and materials reduced lifetime take place.

To minimize these negative effects the stabilizers and antioxidants are most commonly used. However, traditional methods of evaluating the effectiveness of a stabilizer within the PUE require lengthy field tests or the materials exposure to high air temperatures for a period of scores of hours and several days [4].

Modern methods of thermal analysis can significantly reduce the time of polymer tests, and informative results, their accuracy and capability to forecast the coating lifetime are significantly improved and expanded [5–7]. In particular, the determination of oxidation induction time (OIT) and the oxidation onset temperature (OOT) by differential scanning calorimetry (DSC) is effective for the accelerated study of thermal-oxidation polymers stability.

This rapid method has been recommended [8–10] and used for polyolefins [11–15] oils and hydrocarbons [16, 17] and PVC [18].

Information on the use of OIT method for polyurethanes is currently limited. There are only some patent data [19] and publications on the results of determination of OIT and OOT for automotive coating materials, derived from polyurethanes of simple and complex polyester structure [20]. There are actually no publications on the test techniques and results of evaluating the thermal oxidation stability of PUE based on diene oligomers by using DSC. In addition, it should be noted that there are quite a number of manufacturers of commercial stabilizers in the market. Experience has proven that even with the same chemical structure their efficacy may vary. For this reason, when formulating the composition and selecting the stabilizer, or when making a decision on the feasibility and acceptability of direct replacement of one brand by another, one must use a modern rapid method that would quickly assess and predict the thermal stability of the coating material.

In view of the above, the purpose of the present research is to study, by using DSC, the oxidation stability of PUE samples derived from butadiene and isoprene copolymer, and comparative assessment of OIT performance in the presence of different brands of pentaerythritol tetrakis [3-(3,'5'-di-tert-butyl-4'-hydroxyphenyl)propionate] stabilizer.

2.2 EXPERIMENTAL PART

To obtain PUE we used an oligomer, which is a product of the anionic copolymerization of butadiene and isoprene in the ratio of 80:20. The molecular weight of 3200. Mass fraction of hydroxyl groups was 1%, and the oligomer functionality on them was 1.8.

The compositions were being prepared in a ball mill from 12 to 15 h. Homogenization of the components was carried out to the degree of grinding equal to 65. All formulations contained the same amount of the following ingredients: the above oligomer, filler (calcium carbonate), plasticizer of a complex ester nature, desiccant (calcium oxide), organic red pigment (FGR CI 112, produced by Ter Hell & Co Gmbh.).

Stabilizing agents varied in the amounts of 0.2, 0.6, and 1.0 wt. parts to 100 oligomer weight parts. The compositions were numbered in accordance with Table 2.1. Comparison sample was the material under the code 0, which did not contain the stabilizer.

TABLE 2.1   Brands and Content of the Stabilizers Used

PUE sample number	Stabilizer brand and content (per oligomer 100 weight parts)			
	Irganox 1010	Evernox 10	Songnox 1010	Chinox 1010
0	–	–	–	–
1	0.2	–	–	–
2	0.6	–	–	–
3	1	–	–	–
4	–	0.2	–	–
5	–	0.6	–	–
6	–	1	–	–
7	–	–	0.2	–
8	–	–	0.6	–
9	–	–	1	–
10	–	–	–	0.2
11	–	–	–	0.6
12	–	–	–	1

The compositions were cured taking into account the general content of the hydroxyl groups in the system under the action of the estimated volume of Desmodur 44 V20L polyisocyanate (Bayer MaterialScience AG). Mass fraction of isocyanate groups in the product was 32%. Chain branching agent was chemically pure glycerine, and the catalyst was dibutyl tin dilaurate (manufactured by "ACIMA Chemical Industries Limited Inc.).

Curing conditions: standard laboratory temperature and humidity, duration-72 h.

Pentaerythritol tetrakis [3-(3,'5'-di-tert-butyl-4'-hydroxyphenyl)propionate] of Irganox 1010, Evernox 10, Songnox Chinox 1010 and 1010 trademarks was used as a stabilizer. Structural formula is shown in Fig. 2.1.

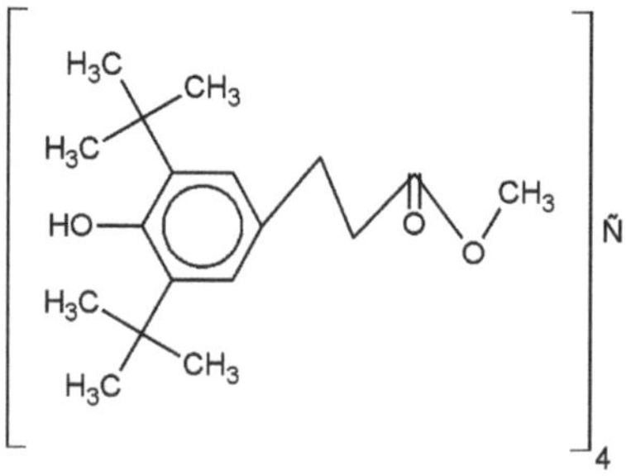

FIGURE 2.1   The structural formula of pentaerythritol tetrakis [3-(3,'5'-di-tert-butyl-4'-hydroxyphenyl)propionate] stabilizer.

Samples of cured PUE were tested by the Netzsch DSC 204 F1 Phoenix heat flow differential scanning calorimeter. Calibration was performed on an indium standard sample. Samples weighing 9 to 12 mg were placed in an open aluminum crucible. Test temperatures were attained at 10K/min rate under constant purging with an inert gas (argon). Upon reaching the target temperature the inert gas supply was stopped and the oxygen supply started at 50 mL/min rate. All data were recorded and processed using Netzsch Proteus special software in OIT registration mode.

2.3   RESULTS AND DISCUSSION

For polyolefins the tests to determine OIT are standardized in terms of both the recommended oxidizing gas (oxygen) flow and the isothermal segment temperature [8, 9]. There are no such standards for PUE. The authors [21] recommend that the test temperature be previously identified experimentally, and other settings be selected in accordance with the recommendations of the relevant ASTM or ISO. For this reason, we first determined the conditions of OIT fixation for the nonstabilized sample at two different temperatures (Fig. 2.2).

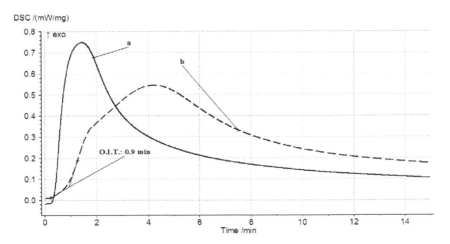

FIGURE 2.2   OIT determination for a nonstabilized PUE (number zero) at 200 °C (a) and 180 °C (b) by using atmosphere oxygen.

As Figure 2.2 shows, isothermal mode at 200 °C does not allow estimating the value of OIT for unstabilized sample. Under these conditions, snowballing oxidation degradation (curve a) starts almost immediately and oxidation induction time cannot be actually defined. As a result of reducing temperature to 180 °C it is possible to fix the target parameter. For unstabilized sample the OIT value was 0.9 min (curve b). However, we found that tests at lower temperatures lead to the significant increase in test duration, especially for stabilized samples. This is undesirable, since the benefit of rapid OIT method in this case is partially lost. Thus, we found the necessary balance between time and temperature conditions that allow estimating OIT for the investigated objects at the recommended oxygen supply rate. In this regard, all subsequent tests were carried out at 180 °C, and the results obtained are illustrated in each case, depending on the stabilizer type and content in comparison with the nonstabilized sample. The sample numbering and the stabilizer amount are consistent with Table 2.1.

Figure 2.3 shows the DSC curves for the samples containing Irganox 1010.

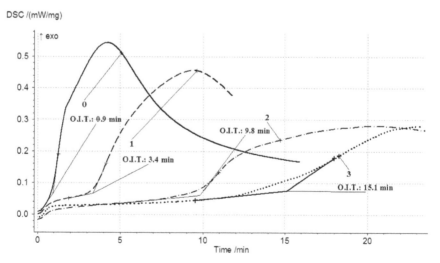

FIGURE 2.3   Isothermal DSC scans at 180°C for samples containing different amounts of Irganox 1010 stabilizer.

On DSC-curves of the materials stabilized by Irganox 1010 antioxidant, the OIT value changes can be traced depending on the content of pentaerythritol tetrakis [3-(3,'5'-di-tert-butyl-4'-hydroxyphenyl)propionate] of this

TABLE 2.2 *(Continued)*

brand. Significant stabilizing effect can be observed even at proportion of 0.2 weight parts. When adding 0.6 and 1.0 weight parts of this product to PUE the OIT was 9.8 and 15.1 min, respectively, which is 10 and 15 times as large as the corresponding parameter of reference sample.

Thereby it should be noted that the detected effects confirm and significantly specify the data obtained earlier [22] by using oligodiendiols and substituted phenol of this particular type, but using the classical evaluation method. The principal difference is that in the latter case, the implementation of the standard involves the need of thermostatic control of samples at higher air temperature (usually within 72 h), the subsequent physical and mechanical testing and correlation of properties before and after thermal aging. To assess the PUE sample oxidation stability by evaluating OIT by means of DSC the time expenditure is no more than 30 min and requires very little sample weight.

Materials containing Evernox 10, Songnox Chinox 1010 and 1010 were also investigated by this method. DSC data are shown in Figs. 2.4–2.6.

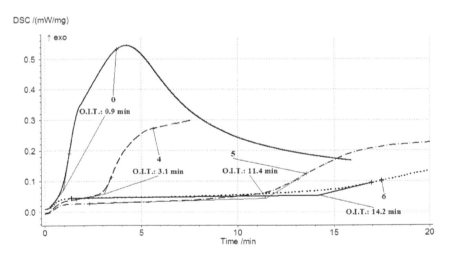

FIGURE 2.4 Isothermal DSC scans at 180 °C for samples containing different amounts of Evernox 10 stabilizer.

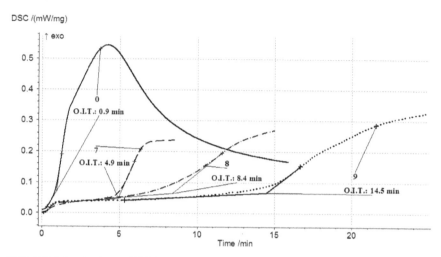

FIGURE 2.5   Isothermal DSC scans at 180°C for samples containing different amounts of Songnox 1010 stabilizer.

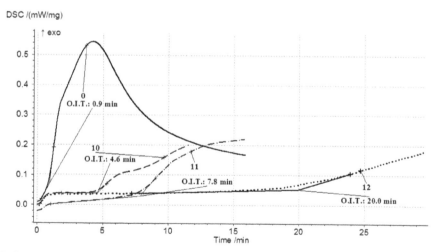

FIGURE 2.6   Isothermal DSC scans at 180°C for samples containing different amounts of Chinox 1010 stabilizer.

For general benchmarking the data obtained by processing the experimental array, are summarized in Table 2.2.

TABLE 2.2 OIT Values for PUE, Depending on the Type and Content of the Stabilizer

Sample number	Stabilizer brand and content (per 100 weight parts)				OIT, min
	Irganox 1010	Evernox 10	Songnox 1010	Chinox 1010	
0	–	–	–	–	0.9
1	0.2	–	–	–	3.4
2	0.6	–	–	–	9.8
3	1	–	–	–	15.1
4	–	0.2	–	–	3.1
5	–	0.6	–	–	11.4
6	–	1	–	–	14.2
7	–	–	0.2	–	4.9
8	–	–	0.6	–	8.4
9	–	–	1	–	14.5
10	–	–	–	0.2	4.6
11	–	–	–	0.6	7.8
12	–	–	–	1	20

It follows from the OIT numerical values that regardless of the stabilizer brand, with an increase of its content in PUE within the investigated concentration range, the natural increase in the oxidation induction time is recorded. However, due to the high sensitivity of the method a significant difference in stabilizing effect of equal quantities of the single-type product manufactured by different vendors can be easily traced. Simple calculation shows that the deviation between the OIT maximum and minimum values for materials stabilized by 0.2 weight parts of Irganox 1010, Evernox 10, Songnox Chinox 1010 and 1010, is equal to 36.7%. With the content being 0.6 and 1.0 weight parts, this deviation is 20.4 and 31.5%, respectively. Apparently, this difference is due to the chemical purity and other factors that determine the protective capacity of the products used.

2.4 CONCLUSIONS

Thus, in the example of mesh polyurethane materials based on copolymer of butadiene and isoprene we show the high efficiency of using the option of

OIT determination by DSC in order to carry out express tests on PUE thermal stability. The preferred temperature of the isothermal segment for accelerated test of this type of materials has been deduced from experiment.

Comparative evaluation of OIT indicators for PUE, stabilized by pentaerythritol tetrakis [3-(3,'5'-di-tert-butyl-4'-hydroxyphenyl)propionate], depending on the manufacturer, ceteris paribus, revealed significant difference in terms of the protective effect of sterically hindered phenol. In practical terms, this means that prior to making the composition and selecting the stabilizer, as well as planning direct qualitative and quantitative replacement of one brand stabilizer by another in the PUE, one must take into consideration the potential significant differences in antioxidant efficacy of the products.

ACKNOWLEDGMENT

This work was supported by the Grant Council of the President of the Russian Federation, grant MK-4559.2013.3.

KEYWORDS

- DSC
- Material testing
- Oxidation induction time
- Polyurethane elastomers
- Stabilizers

REFERENCES

1. Prisacariu, C. (2011). Polyurethane Elastomers. From Morphology to Mechanical Aspects, *Springer-Verlag*, Wien.
2. Novakov, I. A., Nistratov, A. V., Medvedev, V. P., Pyl'nov, D. V., Myachina, E. B., Lukasik, V. A., et al. (2011). Influence of Hardener on Physicochemical and Dynamic Properties of Polyurethanes Based on α, ω-di (2-hydroxypropyl)-Polybutadiene Krasol LBH-3000. *Polymer Science-Series D, 4(2)*, 78–84.
3. Novakov, I. A., Nistratov, A. V., Pyl'nov, D. V., Gugina, S. Y. & Titova, E. N. (2012). Investigation of the Effect of Catalysts on the Foaming Parameters of Compositions and Properties of Elastic Polydieneurethane Foams, *Polymer Science-Series* D, *5(2)*, 92–95.
4. ISO 188:2011 Rubber, Vulcanized or Thermoplastic. Accelerated Ageing and Heat Resistance Tests.
5. Thermal Analysis of Polymers. Fundamentals and Applications. Ed. Joseph D. Menczel, R. Bruce Prime, John Wiley & Sons, Inc, Hoboken, New Jersey, 2009.

6. Principles and Applications of Thermal Analysis. Edited by Paul Gabbott, Blackwell Pub, Oxford, Ames, Iowa 2008.
7. Pieichowski, J. & Pielichowski, K. (1995). Application of Thermal Analysis for the Investigation of Polymer Degradation Processes, *J. Therm. Anal.*, *43*, 505–508.
8. ASTM D 3895–07: Standard Test Method for Oxidative-Induction Time of Polyolefins by Differential Scanning Calorimetry.
9. ISO 11357–6: Differential Scanning Calorimetry (DSC). Determination of Oxidation Induction Time (isothermal OIT) and Oxidation Induction Temperature (dynamic OIT).
10. ASTM E2009–08: Standard Test Method for Oxidation Onset Temperature of Hydrocarbons by *Differential Scanning Calorimetry*.
11. Gomory, I. & Cech, K. (1971). A New Method for Measuring the Induction Period of the Oxidation of Polymers, *J. Therm. Anal.*, *3*, 57–62.
12. Schmid, M., Ritter, A. & Affolter, S. (2006). Determination of Oxidation Induction Time and Temperature by DSC, *J. Therm. Anal. Cal.*, *83–2*, 367–371.
13. Woo, L., Khare, A. R., Sandford, C. L., Ling, M. T. K. & Ding, S. Y. (2001). Relevance of High Temperature Oxidative Stability Testing to Long Term Polymer Durability, *J. Therm. Anal. Cal.*, *64*, 539–548.
14. Peltzer, M. & Jimenez, A. (2009). Determination of Oxidation Parameters by DSC for Polypropylene Stabilized with Hydroxytyrosol (3,4-dihydroxy-phenylethanol), *J. Therm. Anal. Cal.*, *96(1)*, 243–248.
15. Focke Walter, W. & Westhuizen Isbe van der (2010). Oxidation Induction Time and Oxidation Onset Temperature of Polyethylene in Air, *J. Therm. Anal. Cal. 99*, 285–293.
16. Simon, P. & Kolman, L. (2001). DSC Study of Oxidation Induction Periods, *J. Therm. Anal. Cal*, *64*, 813–820.
17. Conceicao Marta, M., Dantas Manoel, B., Rosenhaim Raul, Fernandes, Jr., Valter, J., Santos Ieda, M. G. & Souza Antonio, G. (2009). Evaluation of the Oxidative Induction Time of the Ethylic Castor Biodiesel, *J. Therm. Anal. Cal.*, *97*, 643–646.
18. Woo, L., Ding, S. Y., Ling, M. T. K. & Westphal, S. P. (1997). Study on the Oxidative Induction Test Applied to Medical Polymers, *J. Therm. Anal.*, *49*, 131–138.
19. Dietmar Mäder (Oberursel, DE). (2008). Inventors: Stabilization of Polyol or Polyurethane Compositions Against Thermal Oxidation, US20090137699, USA.
20. Simon, P., Fratricova, M., Schwarzer, P. & Wilde, H.-W. (2006). Evaluation of the Residual Stability of Polyuretane Automotive Coatings by DSC, *J. Therm. Anal. Cal.*, *84(3)*, 679–692.
21. Clauss, M., Andrews, S. M. & Botkin, J. H. (1997). Antioxidant Systems for Stabilization of Flexible Polyurethane Slabstock, *Journal of Cellular Plastics, 33*, 457.
22. Medvedev, V. P., Medvedev, D. V., Navrotskii, V. A. & Lukyanichev, V. V. (2007). The Study of Oxidative Aging Polydieneurethane, *Polyurethane Technologies, 3*, 34–36.

CHAPTER 3

TRENDS IN AROMATIC POLYESTERS

ZINAIDA S. KHASBULATOVA and GENNADY E. ZAIKOV

CONTENTS

ABSTRACT

The data on aromatic polyesters based on phthalic and *n*-oxybenzoic acid derivatives have been presented and various methods of synthesis of such polyesters developed by scientists from different countries for last 50 years have been reviewed.

3.1 INTRODUCTION

The important trend of modern chemistry and technology of polymeric materials is the search for the possibilities of producing materials with novel properties based on given combination of known polymers.

One of the most interesting ways in this direction is the creation of block-copolymers macromolecules of which are the "hybrids" of units differing in chemical structure and composition. Thermodynamical incompatibility of blocks results in stable microphase layering in the majority cases what, finally, allows one to combine the properties of various fragments of macromolecules of block-copolymers in an original way.

Depending on diversity of chemical nature of blocks, their length, number and sequence as well as their ability to crystallize one can obtain materials of structure and properties distinguishing from that of initial components. Here are the huge potential possibilities the practical realization of which has already been started. The most evident is the creation of thermoplastic elastomers (TPEs)–high-tonnage polymeric materials synthesized on the base of principle of block-copolymerization: joining of properties of both thermoplastics and elastomers in one material. The great potentials of block-copolymers caused considerable attention to them within the last years.

Nowadays, all the main problems of physics and physic-chemistry of polymers became closely intertwined when studying block-copolymers: the nature of ordering in polymers, the features of phase separation in polymers and the influence of general molecular parameters on it, the stability of phases at exposing to temperature and power impacts, the features of physical and mechanical properties of microphases and the role of their conjugation.

Existing today numerous methods of synthesis of block-copolymers give the possibility to combine unlimited number of various macromolecules, what already allowed people to synthesize multiple block-copolymers. The thermal and mechanical properties and also the stability of industrial block-copolymers vary in broad limits.

The range of operating temperatures and the thermal stability of TPEs have lately been extended owing to the use of solid blocks of high T_{glass} (of polysulfones or polycarbonates for instance) combined with soft blocks of low T_{glass}. Moreover, the incompatibility of those blocks results in independence of elasticity modulus on temperature in broad temperature range. Applying appropriate selection of chemical nature of blocks one can also improve the other properties.

Some limitations and unresolved issues still exist in areas of synthesis, analysis and characterization of properties and usage of block-copolymers. This is a good stimulus for the intense researches and development of corresponding fields of industry.

The most preferred methods of synthesis of block-copolymers are the three following. The polymerization according to the mechanism of "live" chains with the consecutive addition of monomers has been used in the first one. The second is based on the interaction of two preliminary obtained oligomers with the end functional groups. The third one is the polycondensation of the second block at expense of end group of primarily obtained block of the first monomer. The second and the third methods allow one to use great variety of chemical structures.

So, to produce block-copolymers one can avail numerous reactions allowing one to bind, within the macromolecule, blocks synthesized by means of polycondensation at expense of joining or cycloreversion.

The second method of synthesis of block-copolymers allows one to produce polymers after various combinations of initial compounds, one among which is that monomers able to enter the reaction of condensation are added to oligomers obtained by polycondensation.

Generally, the bifunctional components are used for creating the block-copolymers of $(-AB-)_n$ type. The necessary oligomers could be obtained by means of either condensation reactions or usual polymerization. The end groups of monomer taken in excess are responsible for the chemical nature of process in case of condensation.

Until now, the morphological studies have been performed mainly on block-copolymers containing two chemically different blocks A and B only. One can expect the revelation of quite novel morphological structures for three-block copolymers, including three mutually incompatible units $(ABC)_n$. And there are few references on such polymers.

Russian and foreign scientists remarkably succeeded in both areas of creation of new inflammable, heat-and thermal resistant polycondensation polymers and areas of development of methods of performing polycondensation and studying of the mechanism of reactions grounding the polycondensation processes [1–6].

The reactions of polycondensation are the bases of producing the most important classes of heterochain polymers: polyarylates, polysulfones, polyarylenesterketones, polycarbonates, polyamides and others [7–12].

Non-equilibrium polycondensation can be characterized by a number of advantages among the polycondensation processes. These are the absence of exchange destructive processes, high values of constants of speed of growth of polymeric chain, et cetera. However, some questions of nonequilibrium polycondensation still remain unanswered: the mechanism and basic laws of formation of copolymers when the possibility of combination of positive properties of two, three or more initial monomers can be realized in high-molecular product.

The simple and complex aromatic polyesters, polysulfones and polyaryleneketones possess the complex of valuable properties such as high physic-mechanical and dielectric properties as well as increased thermal stability.

There is a lot of foreign scientific papers devoted to the synthesis and study of copolysulfones based on oligosulfones and polyarylenesterketones based on oligoketones.

Because of importance of the problem of creation of thermo-stable polymers possessing high flame-and thermal resistance accompanied with the good physic-mechanical properties, the study of the regularities of formation of copolyesters and block-copolyesters based of oligosulfoneketones, oligosulfones, oligoketones and oligoformals appeared to be promising depending on the constitution of initial compounds, establishing of interrelations between composition, structure and properties of copolymers.

To improve the basic physic-mechanical parameters and abilities to be reused (in particular, to be dissolved), the synthesis of copolyesters and block-copolyesters has been performed through the stage of formation of oligomers with end reaction-able functional groups.

As the result of performed activities, the oligomers of various chemical compositions have been synthesized: oligosulfones, oligoketones, oligosulfoneketones, oligoformals, and novel aromatic copolyesters and block-copolyesters have been produced.

Obtained copoly and block-copolyestersulfoneketones, as well as polyarylates based of dichloranhydrides of phthalic acids and chloranhydride of 3,5-dibromine-n-oxybenzoic acid and copolyester with groups of terephthaloyl-bis(n-oxybenzoic) acid possess high mechanical and dielectric properties, thermal and fire resistance and also the chemical stability. The regularities of acceptor-catalyst method of polycondensation and high-temperature polycondensation when synthesizing named polymers have been studied and the relations between the composition, structure and properties of polymers obtained have been established. The synthesized here block-copolyesters and copolyesters can find application in various fields of modern industry (automobile, radioelectronic, electrotechnique, Avia, electronic, chemical and others) as thermal resistant construction and layered (film) materials.

3.2   AROMATIC POLYESTERS

Aromatic polyesters are polycondensation organic compounds containing complex ester groups, simple ester links and aromatic fragments within their macromolecule in different combination.

Aromatic polyesters (AP) are thermo-stable polymers; they are thermoplastic products useful for reprocessing into the articles and materials by means of formation methods from solutions and melts.

Mainly, plastics and films are produced from aromatic copolyesters. APs can be used also as lacquers, fibrous binding agents for synthetic paper, membranes, hollow fibers, as additions for semiproducts when obtaining materials based on other polymers.

Many articles based on APs appear in industry. The world production of APs increased from 38 millions of tons in 2004 to 50 in 2008, or on 32%. The polymers referred to the class of constructional plastics are distinguished among APs.

Until now, the technical progress in many fields of industry, especially in engineering industry, instrument production, was circumfused namely by the use of constructional plastics. Such exploitation properties of polymers as durability, thermal stability, electroisolation, antifriction properties, optical transparency and others determine their usage instead of ferrous and nonferrous metals, alloys, wood, ceramics and glass [13]. About 1 ton of polymers replace 5–6 tons of ferrous and nonferrous metals and 3–3,5 tons of wood while the economy of labor expenses reaches 800 man-hours per 1 ton of polymers. About 50% of all polymers used in engineering industry are consumed in electrotechnique and electronics. The 80% of all the production of electrotechnique and up to 95% of that of engineering industry has been produced with help of polymers.

The use of construction plastics allows one to create principally new technology of creation the details, machine knots and devices what provides for the high economical efficiency. The construction polymers are well used by means of modern methods: casting and extrusion to the articles operating in conditions of sign-altering loads at temperatures 100–200 °C.

The modern chemical industry gave constructional thermoplastic materials with lowered consumption of material and weight of the machines, devices, mechanisms, reduced power capacities and labor-intensity when manufacturing and exploiting, increased stint.

Nowadays, the radioelectronics, electrotechnical, Avia, shipbuilding and fields of industry can not develop successively without using the modern progressive polymers such as polyarylates, polysulfones, polyesterketones and others which are perspective construction materials. Only Russian industry involves 50 types of plastics including more than 850 labels and various modifications [14]. As a result, the specific weight of products of engineering areas and several other branches of industry produced with help of plastics grew from 32–35% in 1960 till 85–90% in 1990.

The specific weight of construction plastics among the total world production of plastics reached only 5% in 1975 [15]. But plastics of constructional use prevailed in world production of plastics in 1981–1985 [16].

The introduction of polymers has not only positive effect on the state of already existing traditional areas of industry but also determined the technical progress in rocket and atomic industries, aircraft industry, television, restorative surgery and medicine as a whole et cetera. The world production of construction polymers today is more than 10 million tons.

The thermal and heat-resistant constructional plastics take special place among the polymers. The need for such ones arises from fact that the use of traditional polymeric materials of technological assignment is limited by insufficient working thermal resistance, which is usually less than 103–150 °C.

Two classes of polymers have been used for producing high-tonnage thermally resistant plastics: aromatic polyarylates and polyamides [17]. Starting with these and mutually complementing the properties, materials of high operation properties, and thermo-stable plastics of constructional assignment in the first turn, have been produced.

Some widely used and attractive classes of polymers of constructional assignment are considered in detail in the following sections.

3.3 AROMATIC POLYESTERS OF N-OXYBENZOIC ACID

The n-hydroxybenzoic (n-oxybenzoic) acid has been extensively used at polymeric synthesis for the improvement of thermal stability of polymers for the last years [18, 19]. The aromatic polyester "ECONOL," homopolymer of poly n-oxybenzoic acid, possesses the highest thermal resistant among the all homopolymeric polyethers [20] and attracts attention for the industry.

The poly *n*-oxybenzoic acid is the linear high-crystalline polymer with decomposition temperature in inert environment of 550 °C [20]. Below this point decomposition goes extremely slow. The loss in weight at 460 °C in air is 3% during 1 h of thermal treatment and than number is 1% at 400 °C. The CO, CO_2 and phenol are released when decomposing in vacuum at temperatures 500–565 °C, the coke remnant is of almost polyphenylene-like structure. It is assumed [27], that decomposition starts from breaking of ester bonds. The energy of activation of the process of fission of up to 30% of fly degradation products is 249,5 kJ/mol. The following mechanism for the destruction of polymer has been suggested:

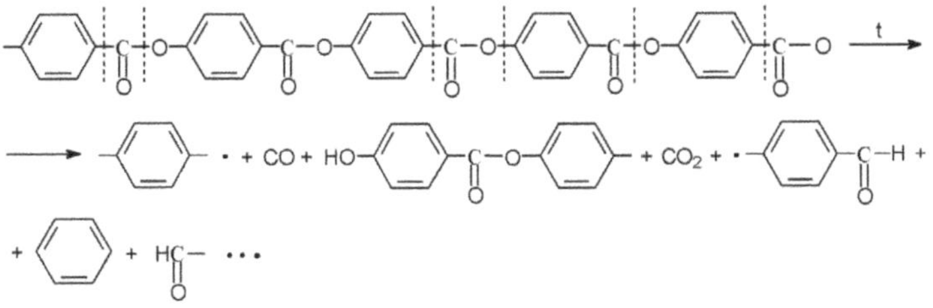

The high ordering of polymer is kept till temperature 425 °C. The equilibrium value of the heat of melting of polymers of *n*-oxybenzoic acid is found to be 5.4 kJ/mol [21].

Several methods for the producing the poly *n*-oxybenzoic acid have been proposed [20, 23–30], which are based of the high-temperature polycondensation, because the phenol hydroxyl possesses low reaction ability. Usually more reactive chloranhydrides are used. The polycondensation of *n*-oxybenzoic acid with blocked hydroxyl group (or of corresponding chloranhydride) happens at temperature of 150 °C and greater. The *n*-acetoxybenzoic acid [23, 24] or the *n*-oxybenzoic acid mixed with acylation agent Ac_2O [25–27] at presence of standard catalyst (in two stages: in solution at temperature 180–280 °C and in solid phase at temperature 300–400 °C) as well as chloranhydride of *n*-oxybenzoic acid [34], or the reaction of *n*-oxybenzoic acid with acidic halogenating agent (for example, $SOCl_2$, PCl_3, PCl_5) [29] have been used at synthesis of high-molecular polymer aiming to increase the reactivity.

Besides the *n*-oxybenzoic acid, its replaced in core derivatives (with F, Cl, Br, J, Me, Et etc. as replacers) can be used as initial materials in polycondensation processes. However, the low-molecular polymer of logarithmic viscosity equal to 0,16 is formed at polycondensation of methyl-replaced *n*-oxybenzoic aced in presence of triphenylphosphite [37], which is associated to the elapsing of adverse reaction of intramolecular etherification with triphenyl phosphite, resulting in termination of chain. This reaction does not flow when using PCl$_3$. High heat-and thermo-stable polyesters of n-oxybenzoic acid of given molecular weight containing no edge COOH-groups can be obtained by heating of n-oxybenzoic acid mixed with dialkylcarbonate at 230–400 °C [20].

The high-molecular polymer can be produced from phenyl ester of n-oxybenzoic acid at presence of butyltitanate in perchloroligophenylene at heating in current nitrogen during 4 h at temperatures 170–190 °C and later after 10 h of exposure to 340–360 °C [20]. Usually [32], Ti, Sn, Pb, Bi, Na, K, Zn or their oxides, salts of acetic, chlorohydratic or benzoic acid are used as catalysts of polycondensation processes when producing polyesters. The polycondensation of polymers of oxybenzoic acids and their derivatives has been performed at absence of catalysts [33] in current nitrogen at temperatures 180–250 °C and pressure during 5–6 h either. The temperature of polycondensation can be lowered [34] if it is carried out in polyphosphoric acid or using activating COOH–group of the substance.

Interesting investigation on study of structure characteristics of polyesters and copolyesters of n-oxybenzoic acid have been performed in Refs. [35, 36]. The introduction of links of n-oxybenzoic acid results in high-molecular compounds of ordered packing of chains, similar to the structure of high-temperature hexagonal modification of homopolymer of n-oxybenzoic acid [37, 38]. The presence of fragments of n-oxybenzoic acid in macromolecular chain of the polymer not only increases the thermal resistance but improves the physic-chemical characteristics of polymeric materials. Obtained polyesters, consisting of monomers of n-oxybenzoic acid and n-dioxyarylene (of formula HO-Ar-OH, where Ar denotes bisphenylene, bisphenylenoxide, bisphenylensulfone), possess higher breaking impact strength, than articles from industrial polyester [39]. The data on thermogravimetric analysis of homopolymer of n-oxybenzoic acid and its polyesters with 4,4'-dioxybisphenylpropane, tere- or isophthalic acids are presented in Ref. [21]. The temperature of 10% loss of mass is 454, 482, 504 °C for them correspondingly.

Highly durable, chemically inert, thermo-stable aromatic polyesters can be produced by interaction of polymers containing links of n-oxybenzoic acid, aromatic dioxy-compounds, for example of hydroquinone and aromatic dicarboxylic acids [40]. Thermo-and chemically resistant polyesters of improved mechanical strength can be obtained by the reaction of n-oxybenzoic acid, aromatic dicarboxylic acids, aromatic dioxy-compounds and diaryl carbonates held in solid phase or in high-boiling solvents [41] at temperature 180 °C and lowered pressure, possibly in the presence of catalysts [42]. Some characteristics of aromatic polyesters based on n-oxybenzoic acid are gathered in Table 3.1 [20].

TABLE 3.1 Physic-Mechanical Properties of Aromatic Polyesters Based on n-Oxybenzoic Acid

Composition of polyesters	Heat resistance on vetch, C, °C	Bending strength, MPa	Flex modulus, N/sm²	Breaking strength, MPa	Elasticity modulus when break, N/sm²
п-oxybenzoic acid, terephthalic acid	–	–	–	35.5	35,000
4,4'-bisphenylquinone				157	47,000
п-oxybenzoic acid, isophthalic acid, hydroquinone, 4,4'-benzophenon dicarboxylic acid	141	163	10,270	–	–
п-oxybenzoic acid, isophthalic acid, hydroquinone, bisphenylcarbonate	–	–	493	120.5	–
п-oxybenzoic acid, isophthalic acid, hydroquinone, 4,4'-bis-hydroxydiphenoxide or bisphenol S	130	160	6100	–	–
п-oxybenzoic acid, isophthalic acid, hydroquinone, 3-chlor-п-oxybenzoic acid and bisphenol D	133	232	10,260	–	–

As mentioned above, the thermal stability of polymers is closely tied to the manifestation of fire-protection properties. Many factors causing the stability of the materials to the exposure of high temperatures are characteristic for the fire-resistant polymers too.

Thermally stable polyesters can be produced on the basis of n-oxybenzoic acid involving stabilizer (triphenylphosphate) introduced on the last stage. The speed of the weight loss decreases two times, after 3 h of exposure to 500 °C polymer loses 0.87% of mass [43].

The high strength high-modular fiber with increased thermo-and fire-resistance can be formed from liquid-crystal copolyester containing 5–95 mol% of n-oxybenzoic acid.

The phosphorus is used as fire-resistant addition, in such a case the oxygen index of the fiber reaches 65% [44].

About 40–70 mol% of the compound of formula Ac-п-C$_6$H$_4$COOH are used to produce complex polyesters of high mechanical strength [45]. The polyesters with improved physic-chemical characteristics can be obtained by single-stage polycondensation of the melt of 30–60 mol% of *n*-oxybenzoic acid mixed with other ingredients [46].

The most used at synthesizing aromatic polyesters are the halogen-replaced anhydrides of dicarboxylic acids and aromatic dioxy-compounds of various constitutions. However, the growing content of halogens in polyesters obtained from mono-, bis-or tetra-replaced terephthalic acid [46] results in lowering of temperatures of glassing, melting as well as of the degree of crystallinity and to some decrease in mechanical strength for halogen-replaced bisphenols. Chlorine-and bromine-containing antipyrenes worsen the thermal resistance of polyesters and are usually used in a company with stabilizers [47, 48] (antipyrenes are chemical substances, either inorganic or organic, containing phosphorus, halogen, nitrogen, boron, metals which are used for the lowering of the flammability of polymeric materials).

Besides, the usage of halogen-replaced bisphenols and dicarboxylic acids for synthesizing polyesters of lowered flammability considerably increases the cost of the latter. Consequently, the aromatic inhibiting flame additions are necessary for creating thermally and fire resistant polymeric materials. The introduction of such agents in small amounts would not diminish properties of polyesters.

Accounting for the aforesaid, one can assume that the chemical modification of known thermally resistant polyesters, by means of introduction of solider component (halogen-containing *n*-oxybenzoic acid) can help to solve the problem of increasing of fire safety of polyesters without worsening of their properties. It is obvious [49] that to replace one should intentionally use bromine atoms which are more effective in conditions of open fire compared to atoms of other halogens.

So, as it comes from the above, the reactive compound involving halogens are widely used for imparting fire-protection properties to aromatic polyesters, as well as oligomer and polymeric antipyrenes.

The liquid-crystal polyesters (LQPs) became quite popular within the last decades. Those differ in their ability to self-arm, possess low coefficient of linear thermal expansion, have extreme size stability, are very chemically stable and almost do not burn.

LQPs can be obtained by means of polycondensation of aromatic oxy-acids (*n*-oxybenzoic one), dicarboxylic acids (iso- and terephthalic ones) and bisphenols (static copolymers) and also by peretherification of polymers and monomers.

Depending on the degree of ordering, LQPs are classified as smectic, nematic and cholesterol ones. Compounds, able to form liquid-crystal state, consist of long flat and quite rigid, in respect to the major axis, molecules.

LQPs can be obtained by several cases:

- polycondensation of dicarboxylic acids with acetylic derivatives of aromatic oxy-acids (*n*-oxybenzoic one) and bisphenols;
- polycondensation of phenyl esters of aromatic oxy-acids (*n*-oxybenzoic one) and aromatic dicarboxylic acids with bisphenols;
- polycondensation of dichloranhydrides of aromatic dicarboxylic acids and bisphenols;
- copolycondensation of dicarboxylic acids, diacetated of bisphenols and/ or acetated of aromatic oxy-acids (*n*-oxybenzoic one) with polyethyleneterephthalate.

Liquid crystal copolyesters have been synthesized [50–56] on the basis of *n*-acetoxybenzoic acid, acetoxybisphenol, terephthalic acid and m-acetoxybenzoic acid by means of polycondensation in melt. All the copolyesters are thermotropic and form nematic phase. The types of LQPs of *n*-oxybenzoic acid are given in Table 3.2.

TABLE 3.2 Label Assortment of Thermotropic Liquid-Crystal Polyesters of *n*-oxybenzoic Acids

Company	Country	Trade label	Remarks
1	2	3	
Dartco Manufacturing Inc.	USA	Xydar	Polyester based on *n*-oxybenzoic acid, terephthalic acid and п, п-bisphenol
		SRT–300	Unfilled with normal fluidity
		SRT–500	Unfilled with high fluidity
		FSR–315	50% of talca
		MD–25	50% of glass fiber
		FC–110	Glass-filled
		FC–120	Glass-filled
		FC–130	Mineral filler
		RC–210	Mineral filler
		RC–220	Glass-filled
LNP Corp.		Thermocomp	Xydar + 150 polytetrafluorethylene (antifriction)

TABLE 3.2   *(Continued)*

Company	Country	Trade label	Remarks
1	2	3	
RTP Co		FDX-65194	Xydar compositions: glass-filed with finishing and thermal-stabilizing additions, mineral fillers
Celanese Corp.		Vectra	Polyesters based on *n*-oxybenzoic acid and naphthalene derivatives
Celanese Corp.		A–130	30% of glass fiber
		B–130	30% of glass fiber
		A–230	30% of carbon fiber
		B–230	30% of glass fiber
		A–540	40% of mineral filler
		A–900	Unfilled
		A–950	Unfilled
Eastman Kodak Co	USA	Vectron	Copolyesters based on *n*-oxybenzoic acid and polyethylene-terephthalate
		LCC–10108	
		LCC–10109	
BASF	Germany	Ultrax	Re-replaced complex aromatic polyester
Bayer A.G.	Germany	Ultrax	Aromatic polyester
ICI	Great Britain	Victrex	Polyester based on *n*-oxybenzoic acid and acetoxynaphtoic acid
		SRP–1500G	Unfilled
		SRP–1500G–30	27% of glass fiber
		SRP–2300G	Unfilled (special design)
		SRP–2300G-30	27% of glass fiber (special design)
Sumimoto kagaku koge K.K	Japan	Ekonol–RE-6000	Polyester based on *n*-oxybenzoic acid, isophthalic acid and bisphenol
Japan Elano Co		Ekonol	Fiber
Mitsubishi chemical Co		Ekonol	Aromatic polyester

TABLE 3.2   *(Continued)*

Company	Country	Trade label	Remarks
1	2	3	
Unitica K.K.		LC–2000	Polyesters based on *n*-oxyben-
		LC–3000	zoic acid and terephthalic acid
		LC-6000	

LQPs can be characterized by high physic-chemical parameters, see Table 3.3.

The fiber possessing high strength properties is formed from the melt of such copolyesters at high temperatures.

The synthesis of totally aromatic thermo-reactive complex copolyesters based on *n*-oxybenzoic acid, terephthalic acid, aromatic diols and alyphatic acids have been performed by means of polycondensation in melt [57–61].

Complex copolyesters are nematic static copolyesters.

The analysis of patents shows that various methods have been proposed for the production of liquid-crystal complex copolyesters [62–64].

The description of methods for production of copolymers (having repetitive links from derivatives of *n*-oxybenzoic acid, 6-oxy-2-naphthoic acid, terephthalic acid and aromatic diol) formable from the melt is reported in papers [65–68]. Each repetitive link is in certain amounts within the polymer. All these copolymers are used as protective covers.

The thermotropic liquid-crystal copolyesters can be synthesized also on the basis of *n*-oxybenzoic and 2,6-oxynaphthoic acid at presence of catalysts (sodium and calcium acetates) [69, 70]. The catalyst sodium acetate accelerates the process of synthesis. Calcium acetate accelerates the process only at high concentration of the catalyst and influences on the morphology of complex copolyesters.

The syntheses of copolymers of *n*-oxybenzoic acid with polyethyleneterephthalate and other components are possible too [71–75]. It is established that copolymers have two-phase nature: if polyethyleneterephthalate is introduced into the reaction mix then copolymers of block-structure are formed.

The properties of liquid-crystal copolyesters based on *n*-oxybenzoic acid and oxynaphthoic acid (in various ratios and temperatures) are described in Refs. [76–81]. It was shown therein that the plates from such polymers formed by die casting are highly anisotropic which results in appearance of the layered structure. Those nematic liquid crystals had ferroelectric ordering. Following substances are characterized in Refs. [82–85]: aromatic copolyesters of *n*-oxybenzoate/bisphenol A; those based on *n*-oxybenzoic acid, bisphenol

TABLE 3.3 Physic-Mechanical Properties of Some Liquid-Crystal Polyesters of *n*-oxybenzoic Acid

Property	Xydar			Vectra				
	SRT-300	SRT-350	FSR-315	A-625 (chem. stable)	A-515 (highly fluid)	A-420 (wear resistant)	A-130 (filled with glass fiber)	C-130 (highly thermal resistant)
Density, g/cm^3	1.35	1.35	1.4	1.54	1.48	1.88	1.57	1.57
Tensile limit, MPa	115.8	125.5	81.4	170	180	140	200	165
Modulus in tension, GPa	9.65	8.27	8.96	10	12	20	17	16
Transverse strength, MPa	131.0	131.0	111.7	–	–	–	–	–
Izod impact, J/m								
Of samples with cut	128.0	208	75	130	370	100	135	120
Of samples without cut	390.0	186	272	–	–	–	–	–
Tensile elongation, %	4.9	4.8	3.3	6.9	4.4	1.3	2.2	1.9
Deformation heat resistance (at 1.8 MPa), °C	–	–	–	185	188	225	230	240
Heat resistance on vetch, °C	366	358	353	–	–	–	–	–
Arc resistance, sec	138	138	–	–	–	–	–	–
Dielectric constant at 10^6 Hz	3.94	3.94	–	–	–	–	–	–
Dielectric loss tangent at 10^6 Hz	0.039	0.039	–	–	–	–	–	–

and terephthalic acid; based on 40 molar% n-oxybenzoic acid, 30 molar% n-hydroquinone and isophthalic acid; and those based on n-oxybenzoic acid, n-hydroquinone and 2,6-naphthalenedicarboxylic acid. The influence of temperature, warm-up time and heating rate on the properties of copolyesters was studied: the glass-transition temperature was shown to increase with warm-up expanded. The number of works is devoted to the study of the complex of physic-chemical properties of liquid-crystal copolyesters based on n-oxybenzoic acid and polyethyleneterephthalate [86–97]. It was established that mixes up to 75% of liquid-crystal compound melt and solidify like pure polyethyleneterephthalate. Copolyesters, containing less than 30% of n-oxybenzoic acid, are in isotropic glassy state while copolyesters of higher concentration of the second compound are in liquid-crystal phase and can be characterized by bigger electric inductivity, than in glassy state. This distinction in caused by the existence of differing orientational distribution of major axes in relation to the direction of electric field in various structure instances of copolyesters. The data of IR-spectroscopy reveal that components of the mix interact in melt by means of reetherification reaction. Increasing pressure, decreasing free volume and mobility in the mix, one can delay the reaction between the compounds.

The investigation of the properties of liquid-crystal copolyesters based on n-oxybenzoic acid/polyethyleneterephthalate and their mixes with isotactic polypropylene (PP), polymethylmethacrylate, polysulfone, polyethylene-2,6-naphthalate, copolyester of n-oxybenzoic acid 6,2-oxynaphthoic acid continues in Refs. [98–106]. Specific volume, thermal expansion coefficient α and compressibility β were measured for PP mixed with liquid-crystal copolymer n-oxybenzoic acid/polyethyleneterephthalate. The high pressure results in appearance of ordering in melted PP. The increase of α and β for all mixes studied (100–25% PP) has been observed at temperatures about melting point of PP having those parameters for liquid-crystal copolyester changed inconsiderably. Increasing content of liquid-crystal copolyester in the mix remarkably decreases the value of α both in solid state and in area of melting PP. The latter has to be accounted for when producing articles. The liquid-crystal compound is grouped in shape of concentric cylinders of various radii in mixes with polysulphones. It becomes responsible for the viscous properties of the mixes at above conditions.

Piezoelectric can be produced from liquid-crystal polymers based on n-oxybenzoic acid/polyethyleneterephthalate and n-oxybenzoic acid/oxynaphthoic acid. Time-stability and temperature range of efficiency of piezoelectric from polymer made of 6,2-oxynaphthoic acid are higher compared to that of polyethyleneterephthalate.

It was found that the use of 30% liquid-crystal copolyester of *n*-oxybenzoic acid/polyethyleneterephthalate makes polymethylmethacrylate 30% more durable and increases the elasticity modulus on 110%, with the reprocessibility being the same.

The structure and properties of various liquid-crystal copolymers are studied in Refs. [107–112]. For example, copolymers based on *n*-oxybenzoic acid, polyethyleneterephthalate, hydroquinone and terephthalic acid, are studied in Ref. [107] the introduction of the mix hydroquinone/terephthalic acid accelerates the process of crystallization and increases the degree of crystallinity of polyesters.

The existence of two structure areas of melts of copolyesters, namely of low-temperature one (where high-melting crystals are present in nematic phase) and high-temperature other (where the homogeneous nematic alloy is formed) is revealed when studying the curves of fluxes of homogeneous and heterogeneous melts of copolyesters based on polyethyleneterephthalate and acetoxybenzoic acid [108].

The curves of flux of liquid-crystal melt are typical for viscoplastic systems while the trend for the flow limit to exist becomes more evident with increasing molar mass and decreasing temperature. Essentially higher quantities of molecular orientation and strength correspond to extrudates obtained from the homogeneous melt compared to extrudates produced from heterophase melt. The influence of high-melting crystallites on the process of disorientation of the structure and on the worsening of durability of extrudates becomes stronger with increasing contribution of mesogenic fragment within the chain.

When studying structure of liquid-crystal copolymers based on *n*-oxybenzoic acid/polyethyleneterephthalate and m-acetoxybenzoic acid by means of IR-spectroscopy, ^{1}H nuclear magnetic resonance and large angle X-ray scattering, the degree of ordering in copolymers was shown [109] to increase if concentration of *n*-oxybenzoic units went from 60 till 75%.

When studying properties of copolymers *n*-oxybenzoate/ ethyleneterephthalate/m-oxybenzoate with help of thermogravimetry [110] the thermo-decomposition of copolymers was found to happen at temperatures 450–457°C in N_2 and 441–447°C in air. The influence of the ratio of *n*-and m-isomers was regarded, the coal yield at $T > 500$ °C found to be 42.6% and increasing with growing number of *n*-oxybenzoic units.

The mixing of melts poly 4,4'-oxybenzoic acid/polyethyleneterephthalate was studied in paper [111] the kinetic characteristics of the mix became incompatible at reetherification.

The measurement of glassing temperature T_{glass} [112] of thermotropic liquid-crystal polyesters synthesized from 4-acetoxybenzoic acid (component A), polyethyleneterephthalate (component B) and 4-acetoxy-hydrofluoric acid revealed that the esters could be characterized by two phases, to which two glassing temperatures correspond: T_{glass} = 66–83°C and T_{glass} = 136–140°C. The lower point belongs to the phase enriched in B, while higher one–to phase enriched in A.

Liquid-crystal polymers are most applicable among novel types of plastics nowadays. The chemical industry is considered to be the one of perspective areas of using liquid-crystal polymers where they can be used, because of their high thermal and corrosion stability, for replacement of stainless steel and ceramics.

Electronics and electrotechnique are considered as promising areas of application of liquid-crystal polymers. In electronics, however, liquid-crystal polymers meet acute competition from cheaper epoxide resins and polysulfone. Another possible areas of using high-heat-resistant liquid crystal polymers are Avia, space and military technique, fiber optics (cover of optical cable and so on), auto industry and film production.

The question of improving fire-resistance of aromatic polyesters is paid more attention last time. Polymeric materials can be classified on criterion of combustibility: noncombustible, hard-to-burn and combustible. Aromatic polyesters enter the combustible group of polymers self-attenuating when taken out of fire.

The considerable fire-resistance of polymeric materials and also the conservation of their form and sizes are required when polymers are exploited to hard conditions such as presence of open fire, oxygen environment, exposure to high-temperature heat fluxes.

On the assumption of placed request, the extensive studies on both syntheses of aromatic polyesters of improved fire-resistance and modification of existing samples of polymers of given type have been carried out. The most used methods of combustibility lowering are following:
- coating by fireproof covers;
- introduction of filler;
- directed synthesis of polymers;
- introduction of antipyrenes;
- chemical modification.

The chemical modification is the widely used, easily manageable method of improving the fire resistance of polymers. It can be done synthetically, simultaneously with the copolymerization with reactive modifier via, for example, introduction of replaced bisphenols, various acids, other oxy-compounds.

Or it can be done by addition of reactive agents during the process of mechanic-chemical treatment or at the stage of reprocessing of polymer melt [113].

The most acceptable ways of modification of aromatic polyesters aimed to get self-attenuating materials of improved resistance to aggressive media are the condensation of polymers from halogen-containing monomers, the combination of aromatic polyesters with halogen-containing compounds and the use of halogen-involving coupling agents [114, 115].

Very different components (aromatic and aliphatic) can become modifiers. They inhibit the processes of combustion and are able to not only to attach some new features to polymers but also improve their physical properties.

3.4   AROMATIC POLYESTERS OF TEREPHTHALOYL-BIS-(N-OXYBENZOIC) ACID

Among known classes of polymers the aromatic polyesters with rigid groups of terephthaloyl-bis-(n-oxybenzoic) acid in main chain of next formula attract considerable attention:

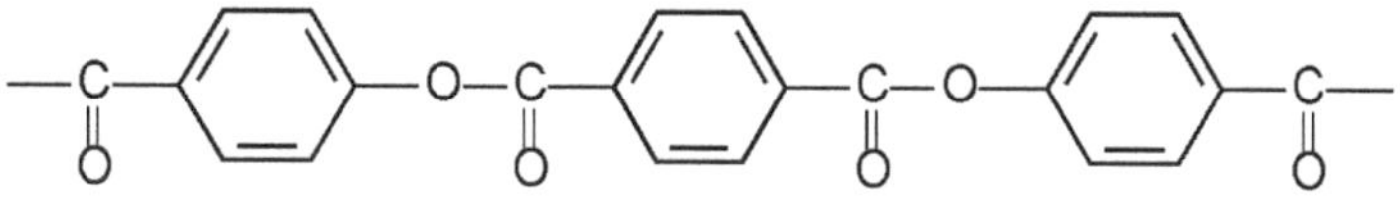

The polymers with alternating terephthaloyl-bis(n-oxybenzoatomic) rigid (R) mesogenic groups and flexible (F) decouplings of various chemical structure within the main chain (RF-copolymers) are able to form liquid-crystal order in the melt. The interest to such polymers is reasoned by fact that dilution of mesogenic "backbone" of macromolecule by flexible decouplings allows one to change the temperature border of polymer transfer from the partially crystalline state into the liquid-crystal one and also to change the interval of existence of liquid-crystal melt. The typical representatives of such class of polymers are polyesters with methylene flexible decouplings.

These polymers have been synthesized by high-temperature polycondensation of terephthaloyl-bis(n-oxybenzoylchloride) with appropriate diols in high-boiling dissolvent under the pressure of inert gas. Bisphenyloxide is used as dissolvent.

The features of conformation, orientation order, molecular dynamics of mentioned class of polymers are studied in Refs. [116–119] on the basis of polydecamethyleneterephthaloyl-bis(n-oxybenzoate), P-10-MTOB, which range of liquid-crystallinity is from 230°C till 290°C.

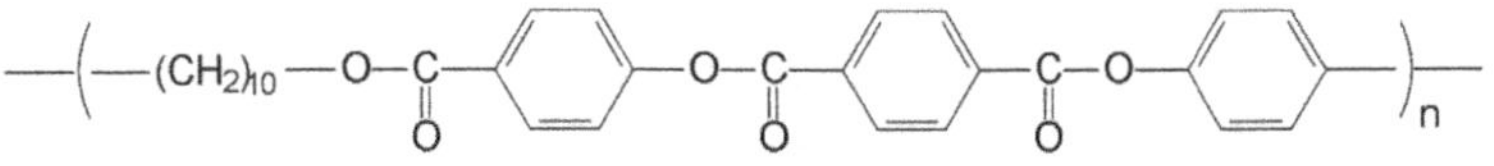

The study of polyesters with terephthaloyl-bis(n-oxybenzoate) groups continues in Refs. [120–123]. The oxyethylene $(CH_2CH_2O)_n$ and oxypropylene $(CH_2CHCH_3O)_n$ groups are used as flexible decouplings. It was found for oxyethylene decouplings that mesophase did not form if flexible segment was 3 times longer than rigid one. The folding of flexible decoupling was found to be responsible for the formation of ordered mesophase in polymers with long flexible decoupling. It was observed that polymers are able to form smectic and nematic liquid-crystal phases, the diapason of existence of which is determined by the length of flexible fragments.

The reported in Ref. [124] were the data on polymers containing methylene siloxane decouplings in their main chain:

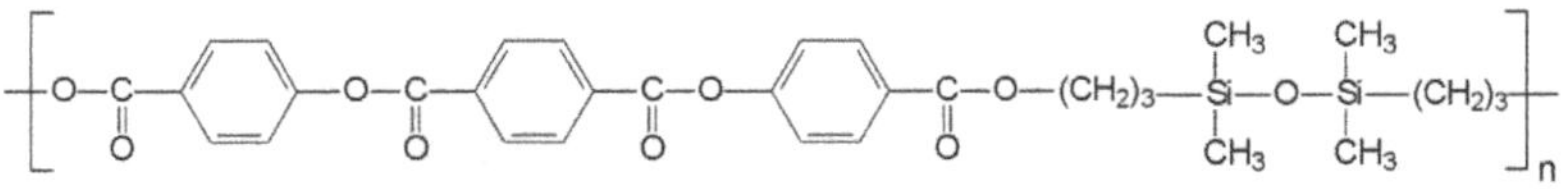

The polymer was produced by heating of dichloranhydride of terephthaloyl-bis(n-oxybenzoic) acid, 1,1,3,3-tetramethylene-1,3-bis-(3-hydroxypropyl)-bis-siloxane and triethylamine in ratio 1:1:2, respectively, in environment of chloroform in argon atmosphere. The studied were the fibers obtained by mechanical extrusion from liquid-crystal melt when heating initial sample. The polymer's characteristic structure was of smectic type with folding location of molecules within layers.

The study of liquid-crystal state was performed in Ref. [125] on polymers with extended (up to five phenylene rings) mesogenic group of various lengths of flexible oxyethylene decouplings:

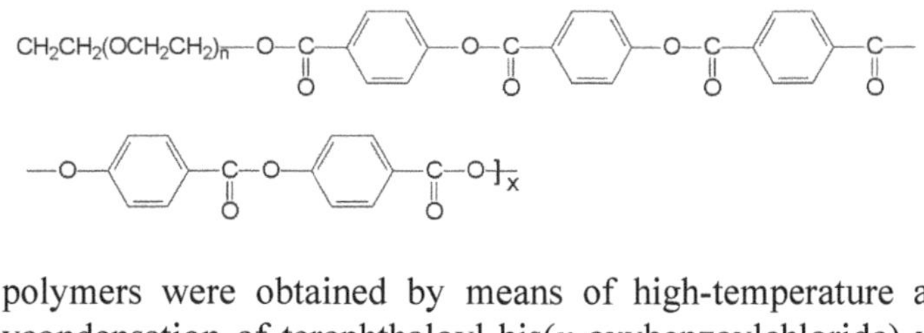

All polymers were obtained by means of high-temperature acceptor-free polycondensation of terephthaloyl-bis(n-oxybenzoylchloride) with bis-4-oxybenzoyl derivatives of corresponding polyethylene glycols in solutions of high-boiling dissolvent in current inert gas. Copolyesters with mesogenic groups, extended up to 5 phenylene cycles and up to 15–17 oxyethylene links, are able to form the structure of nematic type.

Polymers can also form the systems of smectic type [126]. The studied in paper [127] was the mesomorphic structure of polymer with extended group polyethylene glycol-1000-terephthaloyl-bis-4-oxybenzoyl-bis-4'-oxybenzoyl-bis-4''-oxybenzoate:

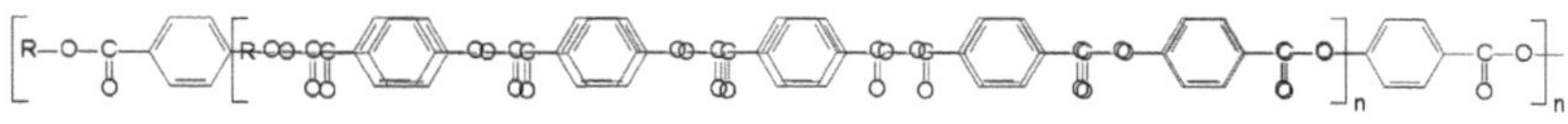

This polymer (of formula R=CH$_2$CH$_2$(OCH$_2$CH$_2$)$_{18-20}$) was synthesized by means of high-temperature acceptor-free polycondensation of terephthaloyl-bis-4-oxybenzoate with bis-(4-oxybenzoyl-4'-oxybenzoyl) derivative poly-ethylene glycol-1000 in the environment of bisphenyloxide [128]. The X-ray diffraction pattern revealed that the regularity of layered order weakened with rising temperature while ordering between mesogenic groups in edge direction kept the same. That type of specific liquid-crystal state in named polymer formed because of melting of layers including flexible oxyethylene decouplings at keeping of the ordering in transversal direction between mesogenic groups.

The Ref. [129] was devoted to the study of the molecular mobility of polymer which formed mesophase of smectic type in temperature range 223–298°C. Selectively deuterated polymers were synthesized to study the dynamics of different fragments of polymer in focus. That polymer was found to have several coexisting types of motion of mesogenic fragment and decoupling at the same temperature. There were massive vibration of phenylene cycles with varying amplitude and theoretically predicted movements of polymethylene chains [130] *trans*-gosh-isomerization and translation motions involving many bonds. Such coexistence was determined by the phase microheterogeneity of polymers investigated.

The process of polycondensation of terephthaloyl-bis(*n*-oxybenzoylchloride) with decamethyleneglycol resulting in liquid-crystal polyester was studied in [131]. The monomers containing groups of complex esters entered into the reaction with diols at sufficiently high temperature (190 °C).

The relatively low contribution of adverse reactions compared to the main one led to regular structure of final polyester. The totally aromatic polyesters of following chemical constitution were studied in [132]: polymer "ФТТ-40" consisting of three monomers (terephthaloyl-bis(*n*-oxybenzoic) acid, terephthalic acid, phenylhydroquinone and resorcinol) entering to the polymeric chain in quantities 2:3:5. Terephthalic acid was replaced with monomer containing phenylene cycle in meta-place in polymer "ФГР-80."

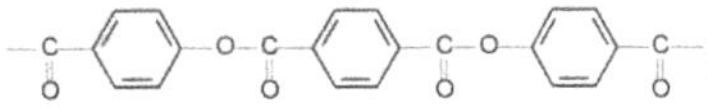

terephthaloyl-bis(n-oxybenzoic) acid (A),

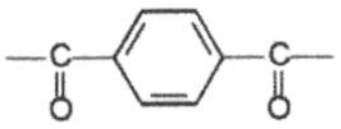

terephthalic acid (B),

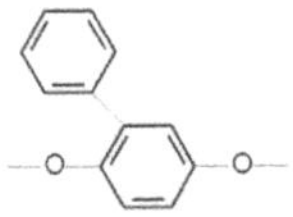

phenylhydroquinone (C),

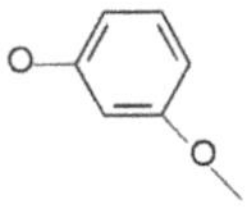

resorcinol (D).

The monomers entered to the polymeric chain in ratio $A:C:D=5:4:1$. These polymers melted at temperatures 300–310°C and switched to liquid-crystal state of nematic type. The monomers in "ФТТ-40" were found to be distributed statistically.

The influence of chemical structure of flexible decouplings in polyterephthaloyl-bis(n-oxybenzoates) on their mesogenic properties was studied in Ref. [131]. The studied were the polymers which had asymmetrical centers introduced into the flexible methylene decoupling and also complex ester groups:

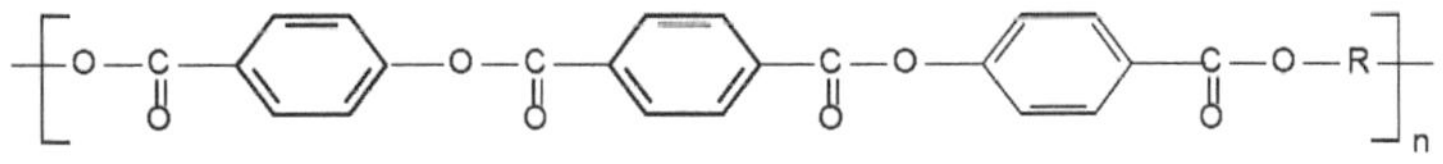

R=CH(CH$_3$)CH$_2$CH$_2$CH$_2$CH$_2$ (I); R=CH(R')COOCH$_2$CH$_2$(II, I)
where R'=CH(CH$_3$)$_2$ (II) or CH$_2$CH$_2$(CH$_3$) (III).

The polymer I was synthesized by means of high-temperature acceptor-free polycondensation from dichloranhydride of terephthaloyl-bis(n-oxybenzoic) acid and 2-methylhexa-methylene-1,5-diol in inert dissolvent (bisphenyloxide at 200 °C). The phase state of polymer I at room temperature was found to depend on the way of sample preparation. The samples obtained from polymer immediately after synthesis and those cooled from melt were in mesomorphic state, but dried from the solutions in trifluoroacetic acid were in partially crystalline state.

The polymers II and III were in partially crystalline state at room temperature irrespective from the method of production.

So, the introduction of asymmetrical center into the flexible pentamethylene decoupling did not change the type of the phase state in the melt and did not influence on the temperature of the transitions into the area of existing of liquid-crystal phase. However, increasing of the rigidity of methylene decouplings via the introduction of complex-ester groups and enlarging of the volume of side branches considerably influenced on the character of intermolecular interaction what resulted in the formation of mesomorphic state of 3D structure during the melting of polymers.

The conformational and optic properties of aromatic copolyesters with links of terephthaloyl-bis(n-oxybenzoic) acid, phenylhydroquinone and resorcinol, containing 5% (from total number of para-aromatic cycles) of m-phenylene cycles within the main chain were studied in Ref. [134]. The polymer had the structure:

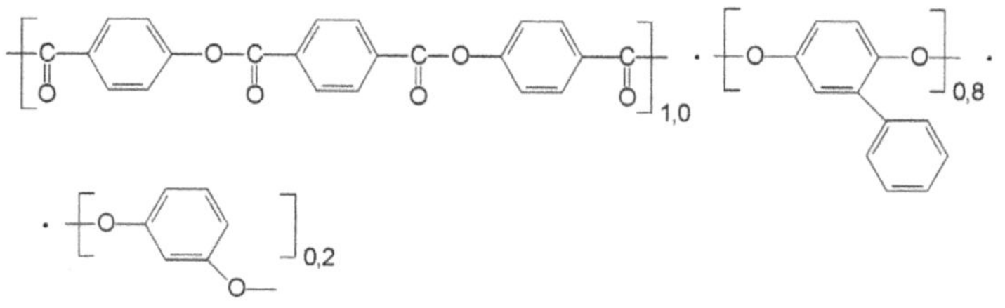

It was determined that the length of statistic Kuhn segment A was $200\pm20\text{Å}$, the degree of dormancy of intramolecular rotations $(\sigma^{-2})^{1}/_{2}=1.08$, the interval of molar masses started with 3.1 and ended 29.9 ′ 10^3.

The analysis of literature data shows that the use of bisphenyl derivatives as elements of structure of polymeric chain allows one to produce liquid crystal polyesters, reprocessible from the melt into the articles of high deformation-strength properties and high heat resistance.

The polyethers containing mesogenic group and various bisphenylene fragments were synthesized in [135]:

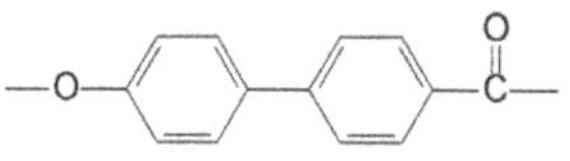 links of 4-oxybisphenyl-4-carbonic acid	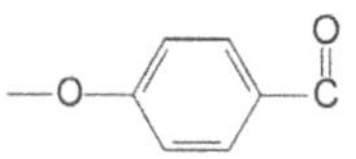 links of n-oxybenzoic acid

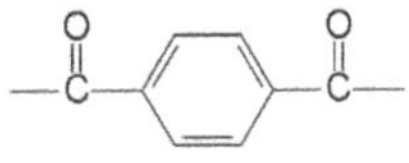 links of terephthalic acid	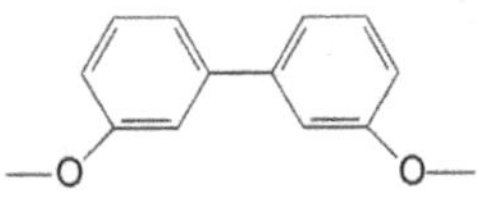 links of dihydroxylic bisphe-nyls of (3,3'-dioxybisphenyl)

All were linked with various decouplings within the main chain of next kinds:

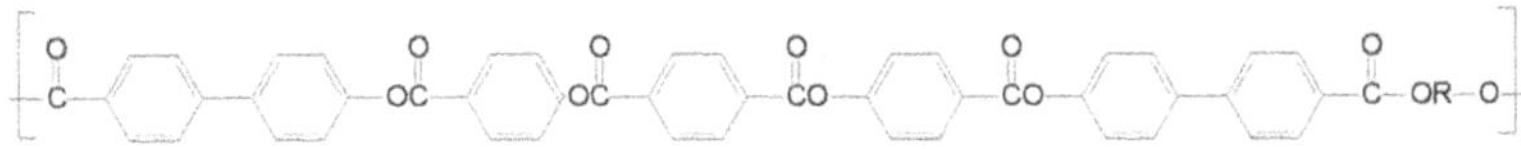

poly(alkyleneterephthaloyl-bis-4-oxybisphenyl-4'-carboxylate)
$[\eta] = 0.78$ dL/g $T_{glass} = 160°C$, $T_{vapor} = 272°C$, where R = $(CH_2)_6$; $(CH_2)_{10}$

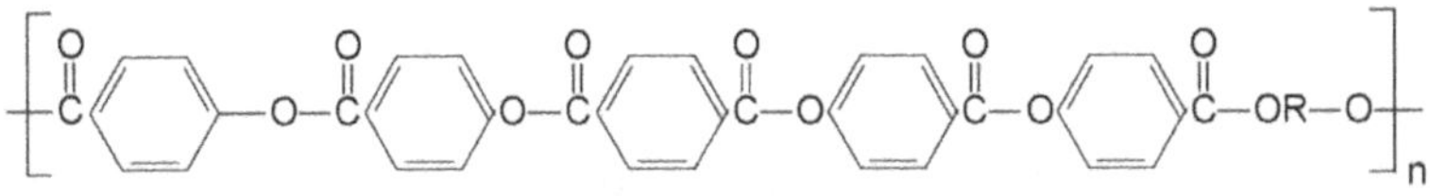

poly(oxyalkyleneterephthaloyl-bis-4-oxybenzoyl-4'-oxybenzoate,
where R = $-CH_2-CH_2OCH_2-CH_2-$; $-(-CH_2CH_2O-)_2-CH_2CH_2-$ oxyethylene decouplings; polyethylene glycols (PEG) PEG 200, PEG 300, PEG 400, PEG 600, PEG 1000–polyoxyethylene decouplings of various molecular weights and copolymers based on 3,3'-dioxybisphenyl with elements of regularity within the main chain:

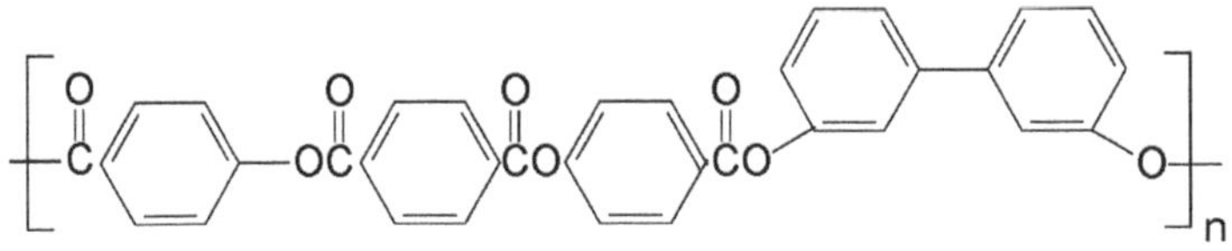

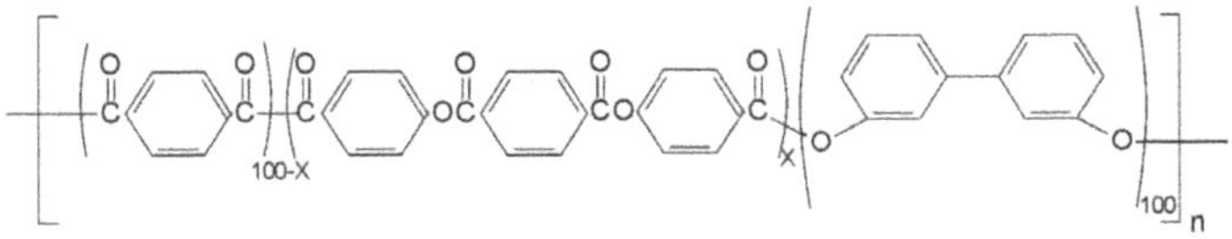

links of terephthaloyl-bis(n-oxybenzoic) acid–links of 3,3'-dioxybisphenyl links of terephthalic acid–links of terephthaloyl-bis(n-oxybenzoic) acid–links of 3,3$'$-dioxybisphenyl.

The synthesized in [136] were the block-copolymers containing the links of terephthaloyl-bis(n-oxybenzoate) and polyarylate. It was found that such polymers had liquid-crystal nematic phase and the biphase separation occurred at temperatures above 280 °C.

The thermotropic liquid-crystal copolyester of polyethyleneterephthalate and terephthaloyl-bis(n-oxybenzoate) was described in Ref. [137]. The snapshots of polarized microscopy revealed that copolyester was nematic liquid-crystal one.

When studying the reaction of reetherification happening in the mix 50:50 of polybutyleneterephthalate and complex polyester, containing mesogenic sections from the remnants of n-oxybenzoic (I) and terephthalic (II) acids separated by the tetramethylene decouplings the content of four triads with central link (I) and three triads with central link II was determined in Ref. [138]. The distribution of triad sequences approximated to the characteristic one for statistical copolymer with increasing of the exposure time.

The relaxation of liquid-crystal thermotropic polymethyleneterephthaloyl-bis(n-oxybenzoate) was studied in Ref. [139]: the β-relaxation occurred at low temperatures while α-relaxation took place at temperatures above 20 °C (β-relaxation is associated with local motion of mesogenic groups while α-relaxation is caused by the glassing in amorphous state). It was established thin the presence of ordering both in crystals and in nematic mesophase expanded relaxation.

So, the liquid-crystal nematic ordering is observed, predominantly, in considered above polymeric systems containing links of terephthaloyl-bis(n-oxybenzoic) acid [140] and combining features of both polymers and liquid crystals. Such ordering can be characterized by the fact that long axes of mesogenic groups are adjusted along some axis and the far translational ordering in distribution of molecules and links is totally absent.

The analysis of references demonstrates that polyesters with links of terephthaloyl-bis(n-oxybenzoic) acid can be characterized by the set of high physic-mechanical and chemical properties.

That is why we have synthesized novel copolymers on the basis of dichloranhydride of terephthaloyl-bis(n-oxybenzoic) acid and various aromatic oligoesters.

3.5   AROMATIC POLYARYLENESTERKETONES

The acceleration of technical progress, the broadening of assortment of chemical production, the increment of the productivity of labor and quality of items from plastics are to a great extent linked to synthesis, mastering and application of new types of polymeric materials. New polymers of improved resistance to air, heat form-stability, longevity appeared to meet the needs of air and space technique. Filling and reinforcement can help to increase the heat resistance on about 100 °C. Plastics enduring the burden in temperature range 150–180°C are called "perspective" while special plastics efficient at 200 °C are named "exotic."

The most promising heat-resistant plastics for hard challenges are polysulfones.

Today, the widely used aromatic polysulfones as constructional and electroisolating materials are those of general formula $[-O-Ar-SO_2-]_n$ where R stands for aromatic radical.

The modified aromatic polysulfones attract greater interest every day. The introduction of ester group into the chain makes the molecule flexible, elastic, more "fluid" and reprocessible. The introduction of bisphenylsulphonic group brings thermal resistance and form stability. Non-modified polysulfones are hardly reprocessible due to the high temperature of melt and are not used for technical purposes.

Aromatic polysulfones are basically linear amorphous polyerylensulfonoxides. They are constructional thermoplasts, which have sulfonic groups-SO_2-in their main chain along with simple ester bonds, aliphatic and aromatic fragments in different combinations.

There are two general methods known for the production of polyarylenesulfonoxides from hydroxyl-containing compounds [8, 141]. These are the polycondensation of disodium or dipotassium salts of bisphenols with dahalogenbisphenylsulfone and the homopolycondensation of sodium or potassium salts of halogenphenoles. One should notice that mentioned reactions occur according to the bimolecular mechanism of nucleophylic replacement of halogen atom within the aromatic core.

The first notion of aromatic polysulfones as promising constructional material was met in 1965 [142–145]. The most pragmatic interest among aromatic polysulfones attracts the product of polycondensation of 2,2-bis(-4-oxyphenyl) of propane and 4,4'-dichlorbisphenyldisulfone:

The degree of polymerization of industrial polysulfones varies from 60 till 120 what corresponds to the weights from 30,000 to 60,000.

The other bisphenols than bisphenylolpropane (diane) can be used for synthesizing polysulfones, and the constitution of ingredients make considerable effect on the properties of polymers.

Linear polysulfones based on bisphenylolpropane and containing isopropylidene groups in the chain are easily reprocessible into the articles and have high hydrolysis stability. The presence of simple ester links in the polymeric chains makes them more flexible and durable. The main effect on properties of such polysulfones is produced by sulfonic bond, which makes the polymer more stable to oxidation and more resistant to heat. Above properties of polysulfones along with the low cost of bisphenylolpropane change them into almost ideal polymers for constructional plastics. Polysulfones of higher heat-resistance can be obtained on the basis of some other bisphenols.

The industrial production of polysulfones based on 2,2-bis-(4oxyphenyl) of propane and 4,4'-dichlorbisphenylsulfone was started by company "Union Carbid" (USA) in 1965 [146].

The manufacture of given aromatic polysulfones was also launched by company "ICI" in Great Britain under the trademark "Udel" in 1966. Various labels of constructional materials (P-1700, P-1700–06, P-1700–13, P-1700–15, etc.) are developed on the basis of this type of polysulfones.

Aromatic polysulfones are soluble in different solvents: good dissolution in chlorized organic solvents and partial dilution in aromatic hydrocarbons.

By its thermo-mechanical properties the aromatic polysulfone on the basis of bisphenylolpropane takes intermediate place between polycarbonate and polyarylate of the same bisphenol. The glassing temperature of the polysulfone lies in the ranges of 190–195°C, heat resistance on vetch is 185 °C. The given polysulfone is devised to be used at temperatures below 150 °C and is frost-resistant material (−100 °C).

One of the most valuable properties of polysulfones is well creep resistance, especially at high temperatures. The polysulfone's creep deformation is 1.5% at 100 °C and after 3.6×10^6 sec loading of 21 MPa, which is better than that of others. Their long-term strength at high temperature is also better. Thus, the polysulfone can be used as constructional material instead of metals.

The flow limit of polysulfone is 71.5 MPa, the permanent strain after rupture is 50–100%. At the same time the elasticity modulus (25.2 MPa) points on sufficient rigidity of material, which is comparable to that of polycarbonate. Impact strength on Izod is 7–8 kJ/m^2 at 23 °C with notch.

The strength and durability of polysulfones keep well at high temperatures. This fact opens possibilities to compete with metals in such areas where other thermoplasts are worthless.

The possibilities of using polysulfones in parts of high-precision articles arise from its low shrinkage (0.7%) and low water absorption (0.22% after 5.64×10^4 sec).

Aromatic polysulfone on the basis of bisphenylolpropane is relatively stable to thermo-oxidation destruction, because the sulfur is in its highest valence state in such polymers; electrons of adjacent benzene nuclei shift, under the presence of sulfur, to the side of sulfogroups what causes the resistance to oxidation.

The polysulfone can operate long at temperature up to 140–170 °C, the loss of weight after 2.52×10^7 sec of loading at 125 °C is less than 0.25%; polymer loses 3% of its mass after 3.24×10^7 sec of exposure to 140 °C in oxygen.

The results of thermal destruction of polysulfone in vacuum have revealed that the first product of decomposition at 400 °C was the sulfur dioxide; there are also methane and bisphenylpropane in products of decomposition at temperatures below 500 °C.

So, the polysulfone is one of the most stable to thermo-oxidation thermoplasts.

The articles from polysulfone possess self-attenuating properties, caused by the nature of polymer but not of the additives. The values of oxygen index for polysulfone lie in the range of 34–38%. Apparently, aromatic polysulfones damp down owing to the formation of carbonized layer, becoming porous protective cover, on their surfaces. The probable is the explanation according to which the inert gas is released from the polymer [8].

Aromatic polysulfone is chemically stable. It is resistant to the effect of mineral acids, alkalis and salt solutions. It is even more stable to carbohydrate oils at higher temperatures and small loadings [149].

The polysulfone on the basis of bisphenylpropane also possesses high dielectric characteristics: volume resistivity–10^{17}; electric inductivity at a frequency 10^6 Hz is 3,1; dielectric loss tangent at a frequency 10^6 Hz is 6 ´ 10^{-4}.

Aromatic polysulfones can be reprocesses by die-casting, by means of extrusion, pressing, blow method [150, 151]. The aromatic polysulfones should be dried out for approximately 1.8 ´ 10^4 sec at 120 °C until the moisture is more than 0,05%. The quality of articles based on polysulfone worsens at higher moists though the polymer itself does not change its properties. The polysulfones are reprocessible at temperatures 315–370 °C.

The high thermal stability of aromatic polysulfones allows one to conduct multitime reprocessing without the destruction of polymer and loss of properties. The pressed procurements can be mechanically treated on common machine tools.

Besides the aromatic polysulfone on the basis of bisphenylpropane, the other aromatic polysulfones are produced industrially. In particular, the industrial manufacture of polysulfones has been carried out by company "Plastik 3 M" (USA) since 1967 (the polymers are produced by reaction of electrophylic replacements at presence of catalyst). The polysulfone labeled "Astrel-360" contains links of following constitution:

The larger amount of first-type links provides for the high glassing temperature (285 °C) of the polymer. The presence of second-type links allows one to reprocess this thermoplast with help of pressing, extrusion and die casting on special equipment [152].

The company "ICI" (Great Britain) produced polysulfones identical to "Astrel-360" in its chemical structure. However, the polymer "720 P" of that firm contains greater number of second-type links. Due to this, the glassing temperature of the polymer is 250 °C and its reprocessing can be done on standard equipment [153].

Pure aromatic structure of given polymers brings them high thermo-oxidation stability and deformation resistance. Mechanical properties of polysulfone allow one to classify them as technical thermoplasts having high durability, rigidity and impact stability. The polymer "Astrel 360" is in the same row with carbon steel, polycarbonate and nylon on its durability.

The polysulfone has following main characteristics: water absorption-1.8% (within 5.64 ´ 10^4 sec); tensile strength – 90 MPa (20°C) and 30 MPa) (260°C); bending strength – 120 MPa (20°C) and 63 MPa (260°C); modulus in tension – 2.8 GPa (20°C); molding shrinkage – 0.8%; electric inductivity at a frequency 60 Hz – 3.94; dielectric loss tangent at 60 Hz – 0.003 [154–157].

The polysulphones have good resistance to the effect of acids, alkalis, engine oils, oil products and aliphatic hydrocarbons.

The polysulfone "Astrel-600" can be reprocesses at harder conditions. Depending on the size and shape of the article, the temperature of pressed material should lie within the ranger of 315–410°C while pressure should be about 350 MPa [158]. The given polysulfone can be reprocessed by any present method. Articles from it can be mechanically treated and welded. Thanks to its properties, polysulfone finds broad use in electronique, electrotechnique and Avia-industry [159, 160].

The Refs. [161–163] reported in 1968 on the production of heat-resistant polyarylenesulfonoxides "Arilon" of following constitution (company "Uniroyal," USA):

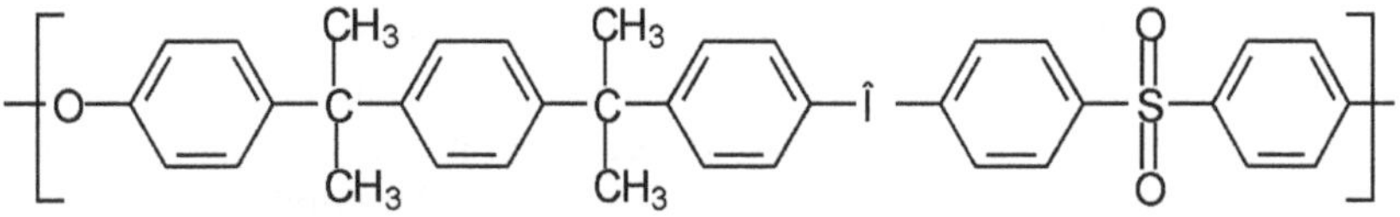

"Arilon" has high rigidity, impact-resistance and chemical stability. The gain weight was 0.9% after 7 days of tests in 20% HCl and 0.5% in 10% NaOH.

This polymer is suitable for long-term use at temperatures 0–130 °C and can be easily reprocessed into articles by die casting. It also can be extruded and exposed to vacuum molding with deep drawing.

Given polysulfone finds application in different fields of technique as constructional material.

Since 1972, the company "ICI" (Great Britain) has been manufacturing the polyarylenesulfonoxide named "Victrex" which forms at homopolycondensation of halogenphenoles [164–166] of coming structure formula:

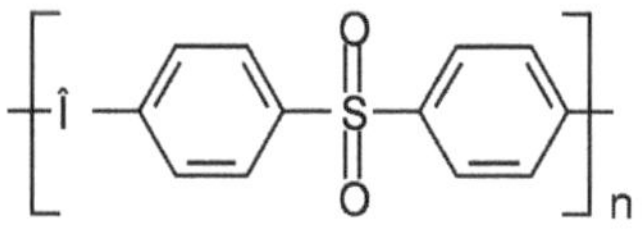

Several types of polysulfone "Victrex" can be distinguished: 100P – powder for solutions and glues; 200P – casting polymer; 300P – with increased molecular weight for extrusion and casting of articles operating under the load at increased temperature in aggressive media; and others.

The polysulfone "Victrex" represents amorphous thermoplastic constructional polymer differing by high heat-resistance, dimensional staunchness, low combustibility, chemical and radiation stability. It can easily be reprocessed on standard equipment at 340–380°C and temperature of press-form 100–150 °C. It is dried at 150 °C for 1.08 ′ 10^4 sec before casting [167, 168].

This polysulfone is out of wide distribution yet. However, it will, presumably, supplant the part of nonferrous metal in automobile industry, in particular, when producing carburetors, oil-filters and others. The polysulfone successively competes with aluminum alloys in Avia industry: the polymer is lighter and yields neither in solidity, neither in other characteristics.

The company "BASF" (Germany) has launched the production of polyethersulfone labeled "Ultrasone" [169] representing amorphous thermoplastic product of polycondensation; it can be characterized by improved chemical stability and fire-resistance. The pressed articles made of it differ in solidity and rigidity at temperature 200 °C. It is assumed to be expedient to use this material when producing articles intended for exposure to increased loadings when the sizes of the article must not alter at temperatures from –100 °C till 150 °C. These items are, for instance in electrotechnique, coils formers, printing and integrated circuits, midspan joints and films for condensers.

"Ultrasone" is used for producing bodies of pilot valves and shaped pieces for hair dryers.

The areas of application of polysulfones are extremely vary and include electrotechnique, car-and aircraft engineering, production of industrial, medical and office equipment, goods of household purpose and packing.

The consumption of polysulfones in Western Europe has reached 50% in electronics/electrotechnique, 23% in transport, 12% in medicine, 7% in space/aviation and 4% in other areas since 1980 till 2006 [170].

The polysulfones are used for manufacturing of printed-circuit substrates, moving parts of relays, coils, clamps, switches, pipes socles, potentiometers details, bodies of tools, alkaline storage and solar batteries, cable and capacitor insulation, sets of television and stereo-apparatuses, radomes. The details under bonnet, the head lights mirrors, the flasks of hydraulic lifting mechanisms of cars are produced from polysulfones. They are also met in internal facing of planes cockpits, protective helmets of pilots and cosmonauts, details of measuring instruments.

The polysulfones are biologically inert and resistant to steam sterilization and γ-radiation what avails people to use them in medicine when implanting artificial lens instead of removed due to a surgical intervention and when producing medical tools and devices (inhalers bodies, ophthalmoscopes, etc.).

The various methods of synthesizing of polysulfones have been devised by Russian and abroad scientists within the last 10–15 years.

The bisphenylolpropane (diane), 4,4'-dioxybisphenylsulfone, 4,4'-dioxybisphenyl, phenolphthalein, hydroquinone, 4,4'-dioxyphenylsulfonyl–bisphenyl are used as initial monomers for the synthesis of aromatic polysulfones. The polycondensation is carried out at temperatures 160–320 °C with dimethylsulfoxide, dimethylacetamide, N-methylpirrolydone, dimethylsulfone and bisphenylsulfone being used as solvents [171–174].

The Japanese researchers report [175–179] on the syntheses of aromatic polysulfones via the method of polycondensation in the environment of polar dissolvent (dimethylformamide, dimethylacetamide, and dimethylsulfoxide) at 60–400 °C in the presence of alkali metal carbonates within 10 min to 100 h. The synthesized thermoplastic polysulfones possess good melt fluidity [175].

The durable thermo-stable polysulfones of high melt fluidity can be obtained by means of polycondensation of the mix of phenols of 1,3-bis(4-hydroxy-1-isopropylidenephenyl and bisphenol A with 4,4'-dichlordimethylsulfone in the presence of anhydrous potassium carbonate in the environment of dimethylformamide at temperature 166 °C. The solution of the polymer is condensed in MeOH, washed by water and dried out at 150 °C in vacuum. The polysulfone has η_{limit} = 0.5 dL/g (1% solution in dimethylformamide at 25°C) [176].

The aromatic polysulfones with the degree of crystallinity of 36% can be synthesized [179] with help of reaction of 2,2-bis(4-hydroxy-4-tret-butylphenyl) propane with 4,4'-dichlorbisphenylsulfone in the presence of potassium carbonate in the environment of polar solvent 1,3-dimethyl-2-imidasolidinone in current nitrogen at temperatures 130–200 °C. After separation and purification polymer has η_{limit} = 0.5 dL/g.

The possibility to use the synthesized aromatic polysulfone as antipyrenes of textile materials was shown in Ref. [180].

The production of polysulfones was shown to be possible in Ref. [181] by means of interaction of sulfuric acid, sulfur trioxide or their mixes with aromatic compounds (naphthalene, methylnaphthalene, methoxynaphthalene, dibenzyl ester, bisphenylcarbonate, bisphenyl, stilbene) if one used the anhydride of carbonic acid at 30–200 °C as the process activator. Obtained polymers could be reprocesses by means of pressing,

The synthesis of polyarylenesulfones containing links of 1,3,5-triphenylbenzene can be performed [182] by oxidation of polyarylene thioesters. Obtained polyarylenesulfones have T_{glass} = 265–329 °C and temperature of 5%

loss of weight of 478–535 °C. They are well soluble in organic solvents and possess fluorescent properties.

To produce the polysulfone with alyphatic main chain, the interaction of SO_2-group of vinyl-aromatic compound (for example of sterol) with non-saturated compound (for instance of acrylonitrile) or cyclic olefin (such as 1,5-cyclooctadiene) has been provided [183] in the presence of initiator of the radical polymerization in bulk or melt, the reaction being carried out at temperatures from 80 to 150 °C.

The method of acceptor-catalyst polyetherification can be used to produce [184, 185] thermo-reactive polyarylenesulfones on the basis of nonsaturated 4,4'-dioxy-3,3'-diallylbisphenyl-2,2'-propane and various chlor-and sulfo-containing monomers and oligomers. The set of physic-mechanical properties of polymers obtained allows one to propose them as constructional polymers, sealing coatings, film materials capable of operating under the influence of aggressive environments and high temperatures.

Relatively few papers are devoted to the synthesis and study of polycon-densational block-copolyesters. At the same time this area represents, un-doubtedly, scientific and practical interest. The number of Refs. [186–190] deals with the problem of synthesis and study of physic-chemical properties of polysulfones of block constitution.

For example, it is reported in Ref. [186] on the synthesis of block-copoly-esters and their properties in dependence on the composition and structure of oligoesters. The oligoesters used were oligoformals on the basis of diane with the degree of condensation 10 and oligosulfones on the basis of phenolphtha-lein with the degree of condensation 10. The synthesis has been performed in conditions of acceptor-catalyst polycondensation.

The polysulfone possesses properties of constructional material: high so-lidity, high thermo-oxidation stability.

However, the deficiency of polysulfonic material is the high viscosity of its melt what results in huge energy expenses at processing. One can decrease the viscosity if, for example, "sews" together polysulfonic blocks with liq-uid-crystal nematic structures which usually have lower values of viscosity. To produce the block-copolymer, the flexible block of polysulfone and rigid polyester block (with liquid-crystal properties) were used in Ref. [187]. The polysulfonic block was used as already ready oligomer of known molecular weight with edge functional groups.

Polyester block represented the product of polycondensation of phenylhy-droquinone and terephthaloylchloride.

The synthesis of block-copolymer went in two stages. At first, the poly-ester block of required molecular weight was produced while on the second stage that block (without separation) reacted to the polysulfonic block. The synthesis was carried out by the method of high-temperature polycondensation in the solution at 250 °C in environment of α-chlornaphthalene. The duration of the first stage was 1.5 h, second–1 h. Obtained polymers exhibited liquid-crystal properties.

The block-copolymers of polysulfone and polyesters can be also produced [188] by the reaction of aromatic polysulfones in environment of dipolar aprotone dissolvents (dimethylsulfoxide, N-methylpirrolydone, N-methyl-caprolactam, N, N{}-dimethylacetamides or their mixes) with alyphatic polyesters containing not less than two edge OH-groups in the presence of basic catalyst–carbonates of alkali metals: Li, Na, K.

Polysulfone is thermo-mechanical and chemically durable thermoplast. But in solution, which is catalyzed by alkali, it becomes sensitive to nucleophylic replacements. In polar aprotone dissolvents at temperatures above 150 °C in the presence of spirit solution of K_2CO_3 it decomposes into the bisphenol A and diarylsulfonic simple esters. The analogous hydrolysis in watery solution of K_2CO_3 goes until phenol products of decomposition. This reaction is of preparative interest for the synthesis of segmented block-copolyester simple polyester–polysulfone. The transetherification of polysulfone, being catalyzed by alkali, results, with exclusion of bisphenol A and introduction segments of simple ester, in formation of segmented block-copolymer [189].

The Refs. [190–199] are devoted to the problem of synthesis of statistical copolymers of polysulfones, production of mixes on the basis of polysulfones and study of their properties as well to the mechanism of copolymerization.

The graft copolymer products, poly(met)acrylates branched to polyester-sulfones, can be produced next way [200]. Firstly, the polyestersulfone is being chlormethylenized by monochlordimethyl ester. The product is used as macrostarter for the graft radical polymerization of methylmethacrylate (I), methylacrylate (II) and butylacrylate (III) in dimethylformamide according to the mechanism of transferring of atoms under the influence of the catalytical system $FeCl_2$/ isophthalic acid. The branched copolymer with I has only one glassing temperature while copolymer with II and III has three.

Lately, several papers on the synthesis of liquid-crystal polysulfones and on the production of mixes and melts of polysulfones with liquid-crystal polymers have appeared [201–205].

For example, the polysulfone "Udel" can be modified [201] by means of introduction of chlormethyl groups. Then the reaction of transquarternization of chlormethylized polysulfone and obtained nitromethylene dimesogens is conducted. The dimesogens contain one phenol OH-group and form nematic mesophase in liquid state. Obtained such a way liquid-crystal polysulfones, having lateral rigid dimesogenic links, possess the structure of enantiotropic nematic mesophase fragments.

The Ref. [202] reports on the synthesis of liquid-crystal polysulfone with mesogenic link–cholesterylpentoatesulfone.

The effect of compatibility, morphology, rheology, mechanical properties of mixes of polysulfones and liquid-crystal polymers are studied in Refs. [203–205]. There are several contributions on the methods of synthesizing of copolymers of polysulfones and polyesterketones and on the production of mixes [206–210]. The method for the synthesis of aromatic copolyestersulfoneketones proposed in Ref. [206] allows one to decrease the number of components used, to lower the demands to the concentration of moist in them and to increase the safety of the process. The method is in the interaction without aseotropoformer in environment of dimethylsulfone of bisphenols, dihaloydarylenesulfones and (or) dihaloydaryleneketones and alkali agents in the shape of crystallohydrated of alkali metal carbonated and bicarbonates. All components used are applicable without preliminary drying.

The novel polyarylenestersulfoneketone containing from the cyclohexane and phthalasynone fragments is produced via the reaction of nucleophylic replacement of 1-methyl-4,5-bis(4-chlorbenzoyl)cyclohexane, 4,4'-dichlorbisphenylsulfone and 4-(3,5-dimethyl-4-hydroxyphenyl)-2,3phthalasine-1-one. The polymer is described by means of IR-spectroscopy with Fourier-transformation, ^{1}H nuclear magnetic resonance, differential scanning calorimetry and diffraction of X-rays. It is shown that the polymer is amorphous and has high glassing temperature (200 °C), it is soluble in several dissolvents at room temperature. The by-products are the "sewn" and graft copolymers [207].

The block copolymers of low-molecular polyesterketoneketone and 4,4' {}-bisphenoxybisphenylsulfone (I) can be produced [208] by the reaction of polycondensation. The glassing temperatures increase and melting points lower of copolymers if the concentration of (I) increases. Block copolymers, containing 32.63–40.7% of (I) have glassing temperature and melting point 185–193 and 322–346°C, respectively; the durability and modulus in tension are 86.6–84.2 MPa and 3.1–3.4 GPa respectively; tensile elongation reaches 18.5–20.3%. Block-copolymers possess good thermal properties and reprocessibility in melt.

The analysis of literature data and patent investigation has revealed that the production of such copolymers as polyestersulfoneketones, liquid-crystal polyestersulfones as well as the preparation of the mixes and melts of polysulfones with liquid-crystal polyesters of certain new properties are of great importance.

With the account for the upper-mentioned, we have synthesized polyestersulfones on the basis of oligosulfones and terephthaloyl-bis(*n*-oxybenzoic) acid [174, 186, 434] and polyestersulfoneketones on the basis of aromatic oligosulfoneketones, mixes of oligosulfones with oligoketones of various constitution and degree of polycondensation and different acidic compounds [211, 212, 435].

3.6 AROMATIC POLYSULFONES

The stormy production development of polymers, containing aromatic cycles like, for example, the polyarylenesterketones has happened within the past decades. The polyarylenesterketones represent the family of polymers in which phenylene rings are connected by the oxygen bridges (simple ester) and carbonic groups (ketones). The polyarylenesterketones include polyesterketone, polyesteresterketone and others distinguishing in the sequence of elements and ratio E/K (of ester groups to ketone ones). This ratio influences on the glassing temperature and melting point: the higher content of ketones increases both temperatures and worsens reprocessibility [213].

The elementary units of polyesteresterketones contain two simple ester and one ketone groups, while those of polyesterketone–only one ester and one ketone [214].

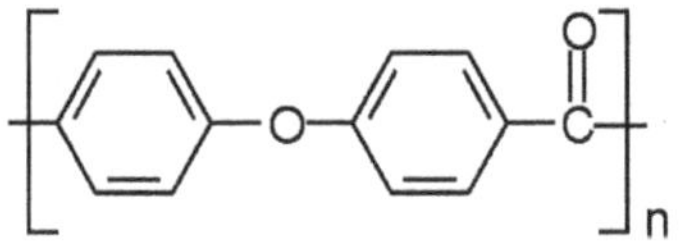

Ї ÝÊ

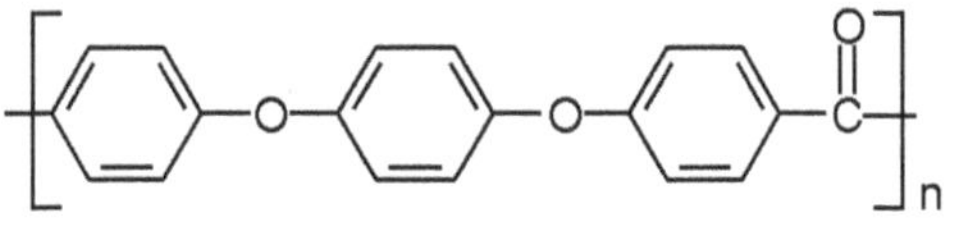

Ї ÝÝÊ

Polyesteresterketone is partially crystalline polymer the thermo-stability of which depends on glassing temperature (amorphosity) and melting point (crystallinity) and increases with immobilization of macromolecules. The strong valence bonds define the high thermo-stability and longevity of mechanical and electrical properties at elevated temperature.

The polyesterketone labeled "Victrex^R" was firstly synthesized in Great Britain by company "Imperial Chemical Industries" in 1977. The industrial manufacture of polyesteresterketones started in 1980 in Western Europe and USA and in 1982 in Japan [215].

The polyesteresterketone became the subject of extensive study from the moment of appearance in industry. The polyesteresterketone possesses the highest melting point among the other high-temperature thermoplasts (335 °C) and can be distinguished by its highly durable and flexible chemical structure. The latter consists of phenylene rings, consecutively joined by para-links to ester, ester and carboxylic groups. There is a lot of information available now on the structure and properties of polyesteresterketones [216, 217].

The polyesteresterketone is specially designed material meeting the stringent requirements from the point of view of heat resistance, inflammability, products combustions and chemical resistance [218, 219]. "Victrex^R" owns the unique combination of properties: thermal characteristics and combustion parameters quite unusual for thermoplastic materials, high stability to effect of different dissolvents and other fluids [220, 221].

The polyesteresterketone can be of two types: simple (unarmored) and reinforced (armored) by glass. Usually both types are opaque though they can become transparent after treatment at certain conditions. This happens due to the reversible change of material's crystallinity which can be recovered by tempering. The limited number of tinges of polyesteresterketones has been produced for those areas of industry where color articles are used.

The structure crystallinity endows polyesteresterketones by such advantages as:
- stability to organic solvents;
- stability to dynamical fatigue;
- improved thermal stability when armoring with glass;
- ability to form plasticity at short-term thermal aging;
- orientation results in high strength fibers.

"Victrex^R" loses its properties as elasticity and solidity moduli with increasing temperature, but the range of working temperatures of polyesteresterketones is wider in short-term process (like purification) than that of other thermoplastic materials. It can be exploited at 300 °C or higher. The presumable stint at 250 °C is more than 50,000 h for the given polyesteresterketone. If one compares the mechanical properties of polyesteresterketone, polyestersulfone, nylon and polypropylene then he finds that the first is the most resistant to wear and to dynamic fatigue. The change of mechanical characteristics was studied in dependence of sorption of CH_2CCl_2.

Polyesteresterketones, similar to any other thermoplastics, is isolation material. It is hard-burnt and forms few smoke and toxic odds in combustion, the demand in such materials arises with all big hardening requirements to accident prevention.

If one compares the smoke-formation in combustion of 2–3 mm of samples from ABC-plastics, polyvinylchloride, polystyrene, polycarbonate, polytetra-fluorethylene, phenolformaldehyde resin, polyestersulfone, polyesteresterketone then it occurs that the least smoke is released by polyesteresterketone, while the greatest amount of smoke is produced by ABC-plastic.

"Victrex[R]" exhibits good resistance to water reagents and pH-factor of different materials starting with 60% sulfuric acid and 50% potassium hydroxide. The polyesteresterketone dissolves only in proton substances (such as concentrated sulfuric acid) or at the temperature close to its melting point. Only α-chlornaphthalene (boiling point 260 °C) and benzophenone influence on "Victrex[R]" among organic dissolvents.

The data on the solubility have revealed that two classes of polyesterester-ketones coexist: "amorphous" and crystalline [222].

The division of these polymers into two mentioned classes is justified only by that the last class, independent of condensation method, crystallizes so fast in conditions of synthesis that the filtering of combustible solution is not possible. It may be concluded from results obtained that "amorphous" class of polyesteresterketones is characterized by bisphenols, which have hybridized sp^3-atom between phenyl groups.

From the point of view of short-term thermal stability the polyesterester-ketones do not yield most steady materials-polyestersulfones-destruction of which is 1% at 430 °C. Yet and still their long-term stability to UV-light, oxygen and heat must be low due to ketone-group [222].

The influence of environment on polyesteresterketones is not understood in detail, but it has proven that polyesteresterketone fully keeps all its properties within 1 year. Polyesteresterketones exhibit very good stability to X-ray, β-and γ-radiations. Wire samples densely covered with polyesteresterketones bear the radiation 110 Mrad without essential destruction.

The destroying tensile stress of polyesteresterketone is almost nil at exposure to air during 100 h at 270 °C. At the same time the flex modulus at glassing temperature of 113 °C falls off precipitously, however remains sufficiently high compared to that of other thermoplasts.

When placed into hot water (80 °C) for 800 h the tensile stress and the permanent strain after rupture of polyesteresterketones decreases negligibly. The polyesteresterketone overcomes all the other thermoplasts on stability to steam action. The articles from polyesteresterketone can stand short exposure to steam at 300 °C.

On fire-resistance this polymer is related to hard-to-burn materials.

The chemical stability of polyesteresterketone "Victrex[R]" is about the same as of polytetrafluorethylene while its long-term strength and impact toughness are essentially higher than those or nylon A-10 [223].

The manufacture of the polyesteresterketones in Japan is organized by companies "Mitsui Toatsu Chem" under the labels "Talpa-2000", "ICI Japan," "Sumitoma Kogaku Koge." The Japanese polyesteresterketones have glassing temperature 143 °C and melting point 334 °C [224, 225].

The consumption of polyesteresterketones in Japan in 1984 was 20 tons, 1 kilogram cost 17000 Ian. The total consumption of polyestersulfones and polyesteresterketones in Japan in 1990 was 450–500 tons per year [226].

Today, 35% of polyesteresterketones produced in Japan (of general formula $[-OC_6H_4-O-C_6H_4-CO-C_6H_4-]_n$) are used in electronics and electrotechnique, 25%-in aviation and aerospace technologies, 10%-in car manufacture, 15%-in chemical industry as well as in the fields of everyday life, for example at producing the buckets for hot water, operating under pressure and temperature up to 300 °C. The Japanese industrial labels of polyesteresterketones have good physical-mechanical characteristics: the high impact toughness; the heat resistance (152 °C, and 286 °C with introduction of 20% of glass fiber); the chemical resistance (bear the influence of acids and alkalis, different chemicals and medicines), the tolerance for radiation action; the elasticity modulus 250–300 kg/mm^2; the rigidity; the lengthening of 100%; the negligible quantity of smoke produced. The polyesteresterketone can be reprocessed by die casting at 300–380 °C (1000–1400 kg/cm^2), extrusion, formation and others methods [224]. High physical-mechanical properties remain the dame for the long time and decrease on 50% only after 10 years.

The polymer is expensive. One of the ways for reducing the product price is compounding. The company "Kogaku Sumitoma" created compounds "Sumiploy K" on the basis of polyesteresterketones by their original technology. The series "Sumiploy K" includes the polymers with high strength and of improved wear resistance. The series "Sumiploy SK" is based on the polyesteresterketone alloyed with other polymers [224]. The series includes polymers, the articles of which can be easily taken out from the form, and products are of improved wear resistance, increased high strength, good antistatic properties, can be easily metalized.

The company "Hoechst" (Germany) produces unreinforced polyesteresterketones "Hoechst X915," armored with 30 wt.% of glass fiber (X925) and carbon fiber (X935), which are characterized by the good physical-mechanical properties (Table 3.4). The unreinforced polymer has the density almost constant till glassing temperature (about 160 °C). The reinforcing with fibers permits further to increase the heat resistance of polyesteresterketone. At the moment the polyesteresterketones labeled "Hostatec" are being produced with 10, 20 and 30 wt.% of glass and carbon fibers. Several labels of polyesteresterketones are under development which involve mineral fillers, are not reinforced and contain 30 wt.% of glass and carbon fibers.

TABLE 3.4 Physic-Mechanical Properties of Polyesteresterketones of Company "Hoechst"

Property	Unreinforced	Containing 30 wt.%	
		Glass fiber	Carbon fiber
Density, g/cm^3	1.3	1.55	1.45
Linear shrinkage, %	1.5	0.5	0.1
Tensile strength, N/mm^2	86	168	218
Breaking elongation, %	3.6	2.2	2.0
Modulus in tension, kN/mm^2	4	13.5	22.5
Impact toughness notched, J/m	51	71	60
Heat resistance, °C	160	Above 320	320

The constructional thermoplast "Hostatec" dominates polyoxymethylene, polyamide and complex polyesters on many parameters.

The polyesteresterketones can be easily processed by pressing, die cast and extrusion. They can be repeatedly crushed to powder for secondary utilization. They are mainly used as constructional materials but also can be used as electroinsulation covers operating at temperatures of 200 °C and higher for a long time [225].

The polyesteresterketone found application in household goods (in this case its high heat resistance and impact toughness are being used), in lorries (joint washers, bearings, probes bodies, coils and other details, contacting fuel, lubricant and cooling fluid).

The big attention to polyesteresterketones is paid in aircraft and space industries. The requirements to fire-resistance of plastics used in crafts have become stricter within the last years. Unreinforced polyesteresterketones satisfy these demands having the fire-resistance category U-O on UL 94 at thickness 0.8 mm. In addition, this polymer releases few smoke and toxic substances in combustion (is used in a subway). The polyesteresterketone is used for coating of wires and cables, used in details of aerospace facility (the low inflammability, the excellent permeability and the wear stability), in military facility, ship building, on nuclear power plants (resists the radiation of about 1000 Mrad and temperature of water steam 185 °C), in oil wells (pillar stand to the action of water under pressure, at a temperature of 288 °C), in electrical engineering and electronics. Polyesteresterketone offer properties of thermoreactive resin, can be easily pressed, undergoes overtone, resists the influence of alkalis.

Since January 1990, the Federal aviation authority have adopted the developed at Ohio State University method, in which the heat radiation (HR) and the rate of heat release (RHR) are determined. The standard regulates the HR level at 65 kJ. Many aircraft materials do not face this demand: for example, the ternary copolymer of acrylonitrile, butadiene and styrene resin, the polycarbonate, phenol and epoxide resins. It is foreseen to substitute these materials by such polymers, which meet these requirements. The measurements in combustion chamber of Ohio State University indicated that polyesteresterketone fulfills this standard.

The heightened activity in creating and evaluation of composite properties on the basis of polyesteresterketones occurs recent years [228]. The thermoplastic composites have the number of advantages regarding the plasticity, maintainability and ability for secondary utilization compared to epoxy composites. These polymers are intended to be applied in alleviated support elements. In the area of cable insulation, polyesteresterketones can be reasonably used when thermo-stability combined with fire-resistance without using of halogen antipyrenes where is desired.

The "Hostatec" has low water absorption. Dielectric properties of films from polyesteresterketone "Hostatec" are high. This amorphous polymer has the electric inductivity 3,6, loss factor 10^{-3} and the specific volume resistance 10^{17} Ohm×cm, these values remain still up to 60 °C.

Growth of demand for polyesteresterketones is very intense. In connection with growing demands on heat resistance and stability to various external factors the polyesteresterketones find broader distribution. The cost of one kilogram of such polymer is 5–20 times larger than the cost of usual constructional polymers (polycarbonates, polyamides, and polyformaldehydes). But, despite the high prices the polyesteresterketones and compositions on their basis, owing to the high level of consumer characteristics, find more and more applications in all industries. The growth of production volumes is observed every year.

It is known, that the paces of annual growth of polyesteresterketones consumption were about 25% before 1995, and its global consumption in 1995 was 4000 tons. The polyesteresterketone do not cause ecological problems and is amenable to secondary processing.

In connection with big perspectiveness of polyarylesterketones, the examining of the most popular methods for their production was of interest. The literature data analysis shows that the synthesis of aromatic polyesterketones can be done by the acylation on Friedel-Crafts reaction or by reaction of nucleophylic substitution of activated dihalogen-containing aromatic compounds and bisphenolates of alkali metals [224, 229].

In the majority cases, the polyesterketones and polyesteresterketones are produced by means of polycondensation interaction of bisphenols with 4,4'-dihalogen-substituted derivatives of benzophenone [230–263], generally it is 4,4'-difluoro-or dichlorbisphenylketone. The introduction of replacers into the benzene ring of initial monomer raises the solubility of polyesterketones and polyesteresterketones. So, polyesterketone on the basis of 3,3',5,5'-tetramethylene and 4,4'-difluorobenzophenone is dissolved at 25 °C [211–214] in dimethylsulfoxide, the reduced viscosity of melt with concentration 0.5 g/dL is 0.79 dL/g.

The polyesterketone based on 4,4'-dioxybenzophenone and dichlormethylenized benzene derivatives is soluble in chloroform and dichloroethane [264]. The logarithmic solution viscosity of polyesterketone obtained on the basis of 4,4'-dioxybisphenylsulfone and 4,4'-dichlorbenzophenone in tetrachloroethane of concentration 0,5 g/dL is 0,486; the film materials from this polyester (of thickness 1 mm) are characterized by the high light transmission (86%) and keep perfect solubility and initial viscosity after exposure to 320 °C during 2 h.

The high-boiling polar organic solvents–dimethylsulfoxide, sulfolane, dimethylsulfone, dimethylformamide, dimethylacetamide–are generally used for synthesizing the polyesterketones and polyesteresterketones by means of polycondensation; in this case the reaction catalysts are the anhydrous hydroxides, carbonates, fluorides and hydrides of alkali metals. The polymers synthesis is recommended to be carried out in inert gas atmosphere at temperatures 50–450 °C. If catalysts used are the salts of carbonic or hydrofluoric acids then oligomers appear. Chain length regulators when producing polyesterketones based on difluoro-or dichlorbenzophenone and bisphenolates of alkali metals and dihydroxynaphthalenes can be the monatomic phenols [251–253].

The synthesis of polyesterketones and polyesteresterketones according to the Friedel-Crafts reaction is lead in mild conditions [265–283]. So, solidifying thermo-stable aromatic polesterketonesulfones, applied as binding agents when laminating, can be produced [278] in the presence of aluminum chloride by the interaction of 1,4-di(4-benzoylchloride)butadiene-1,3-dichloranhydrides of iso-and terephthalic acids, bisphenyl oxide and 4,4'-bisphenoxy-bisphenylsulfone.

The aromatic polyesterketones and their thioanalogs are synthesized [266–281, 284] with help of polycondensation of substituted and not-substituted aromatic esters and thioesters with choric anhydrides of dicarboxylic acids in environment of aprotone dissolvents at temperatures from −10°C till 100 °C in the presence of Lewis acids and bases.

The aromatic polyesterketones and polyesterketonesulfonamides based on 4,4'-dichloranhydride of bisphenyloxidebicarbonic acid and 4-phenoxybenzoylchloride can be produced by means of Friedel-Crafts polycondensation in the presence of $AlCl_3$ [270]. The reduced viscosity of the solution in sulfuric acid of concentration 0.5 g/dL is 0.07–1.98 dL/g. The Friedel-Crafts reaction can also be applied to synthesize the copolyesterketones from bisphenyl ester and aromatic dicarboxylic acids or their halogenanhydrides [274]. The molecular mass of polymers, assessed on the parameter of melt fluidity, peaks when bisphenyl ester is used in abundant amount (2–8%).

The polyarylesterketones can be produced by means of interaction between bisphenylsulfide, dibenzofurane and bisphenyloxide with monomers of electrophylic nature (phosgene, terephthaloylchloride) or using homopolycondensation of 4-phenoxybenzoylchloride and 4-phenoxy-4-chlorcarbonyl-bisphenyl in the presence of dichloroethane at 25 °C [282–285]. Aromatic polyesterketones form after the polycondensation of 4-phenoxybenzoylchloride with chloranhydrides of tere-and isophthalic acids, 4,4'-dicarboxybisphenyloxide in the environment of nitrobenzene, methylchloride and dichloroethane at temperatures from −70°C till 40 °C during 16–26 h according Friedel-Crafts reaction.

The synthesis of polyesterketones based on aromatic ether acids is possible in the environment of trifluoromethanesulfonic acid [249, 287]. The data of ^{13}C nuclear magnetic resonance have revealed [249] that such polyesterketone comprise only the n-substituted benzene rings. When using the N-cyclohexyl-2-pyrrolidone as a solvent when synthesizing polyphenylenesterketones and polyphenylenethioesterketones the speed of polycondensation and the molecular mass of polymers [288] increase.

So, the polymer is produced, the reduced viscosity of 0.5% solution of which in sulfuric acid is 1.0 dL/g, after the interaction of 4,4'-difluorobenzophenone and hydroquinone at 290 °C in the presence of potassium carbonate during 1 h. Almost the same results are obtained when synthesizing polyphenylenethioesterketones. However, the application of the mix bisphenylsulfone/sulfolane as a solvent during the interaction of 4,4'-difluorobenzophenone with sodium sulfide within 2–13 h provides for the production of polymer with reduced viscosity 0.23–0.25 dL/g. The high-molecular polyarylenesulfideketones, suitable for the preparation of films, fibers and composite materials are formed during the interaction of 4,4'-dihalogenbenzophenone with sodium hydrosulfide in solution (N-methylpirrolydone) at 175–350 °C within 1–72 h [289].

For the purpose of improving physical-mechanical properties and increasing the reprocessibility of polyesterketones and polyesteresterketones their sulfonation by liquid oxide of hexavalent sulfur in environment of dichloroethane has been carried out [290]. In this case the polymer destruction does not happen, which is observed at sulfonation by concentrated sulfuric or chlorosulfonic acid. The abiding flexible film materials can be produced from sulfonated materials using the method of casting. The aromatic polyesterketones could also be produced by oxidative dihydropolycondensation (according to the Scolla reaction) of 4,4'-bis-(1-naphthoxy)-benzophenone at 20 °C in the presence of trivalent iron chloride in environment of nitrobenzene [291]. The mix of pentavalent phosphorus oxide and methylphosphonic acid in ratio 1:10 can be used as a solvent and dehydrate agent [292, 293] when synthesizing the aromatic polyesterketones on the basis of bisphenylester of hydroquinone, 4,4'-bisphenyloxybicarbonic acid, 1,4-bis(m-carboxyphenoxy) benzene, and also for homopolycondensation of 3-or 4-phenoxybenzoic acid at 80–140 °C.

The sulfur-containing analogs of polyesterketones, that is, polythioesterketones and copolythioesterketones can be synthesized [294–303] by polycondensation of dihalogenbenzophenols with hydrothiophenol or other bifunctional sulfur-involving compounds, and also of their mixes with different bisphenols in environment of polar organic solvents. As in case of polyesterketones, the synthesis of their thioanalogs is recommended to be carried to out in inert medium at temperatures below 400 °C in the presence of catalyst (hydroxides, carbonates and hydrocarbonates of alkali metals).

The aromatic polyesterketones can also be produced by polycondensation or homopolycondensation of compounds like halogen-containing arylketonephenols and arylenedihalogenides of different functionality at elevated temperature and when the salts of alkali and alkali-earth metals in environments of high-boiling polar organic solvents are used as catalysts [222, 304–312]. The 4-halogen-3-phenyl-4-hydroxybenzophenone, 4-(n-haloidbenzoyl)-2,6-dinethylphenol and others belong to monomers with mixed functional groups.

The polycondensation process when synthesizing polyesterketones and polyesteresterketones can be performed in the melt [223, 313–317] too. So, it is possible to produce the aromatic polyesterketones by means of interaction in melt of 4,4'-difluorobenzophenone with trimethylsiloxane esters of bisphenols with different bridged groups in the presence of catalyst (cesium fluoride–0.1% from total weight of both monomers) at 220–270 °C [318]. The monomers do not enter the reaction without catalyst at temperatures below 350 °C. The reduced viscosity of 2% solution of polymer in tetrachloroethane at 30 °C is 0.13–1.13 dL/g, the molecular weight is 3200–60,000, glassing temperature is 151–186 °C, melting point is 240–420 °C. According to the data of thermogravimetric analysis in the air the mass loss of polymer is less than 10%, when temperature elevates from 422 °C to 544 °C with the rate 8 deg/min.

To increase the basic physical-mechanical characteristics and reprocessibility (the solubility, in particular), the polyesterketones and polyesteresterketones are synthesized [319–329] through the stage of formation of oligomers with end functional groups accompanied with the consequent production of block-copolyesterketones or by means of one-go-copolycondensation of initial monomers with production of copolyesterketones.

The polyesteresterketones, their copolymers and mixes are used for casting of thermally loaded parts of moving transport, instrument, machines, and planes. They are used in articles of space equipment: for cable insulation, facings (pouring) elements. For instance, the illuminators frames of planes and rings for high-frequency cable are made of polyesterketone "Ultrapek." It is widely used in electronics, electrician, for extrusion of tubes and pipes operating in aggressive media and at low temperatures. The polyesteresterketones and polyesterketones are used for multilayer coating as the basis of printed planes. The conservation of mechanical strength in conditions of high humidity and temperature, the stability to radiation forwards their application to aerospace engineering.

The compositions on the basis of polyesteresterketones already compete with those based on thermoreactive resins when making parts of military and civilian aircrafts. Sometimes it is possible to cut the weight on 30% if produce separate parts of plane engines from reinforced polyesteresterketones.

It is proposed to use the polyesteresterketones in the manufacture of fingers of control rod and cams of brake system, motors buttons in car industry. The piston cap of automobile engine, made from polyesteresterketone "Victrex," went through 1300 h of on-the-road tests.

The advantage of using this material, contrary to steel, lays in wearout decrease, noise reduction, 40% weight reduction of the article. The important area of application of polyesteresterketones can be the production of bearings and backings. The polyesteresterketones are recommended to be used for piece making of drilling equipment (zero and supporting mantles) and timber technique, in different joining's of electric equipment of nuclear reactors, layers and valves coverings, components of sports facility.

The fibers of diameter 0.4 mm (from which the fabric is weaved in shape of tapes and belts used in industrial processes where the temperature-resistant, the high-speed conveyers are needed) are produced from polyesteresterketone melt at temperatures of 350–390 °C. The fabric from polyesteresterketone or polyesterketone keeps 90% of the tensile strength after thermal treatment at 260 °C, does not change its properties after steaming at 126 °C for 72 h under the load, and resists alkalis action with marginal change. The medical instruments, analytical, dialysis devices, endoscopes, surgical and dental tools, containers made of polyarylesterketones can be sterilized by steam and irradiation [330].

The manifold methods of synthesizing polyarylenesterketones have been devised by Russian and foreign scientists within the last 20 years.

For example, the method for production of aromatic polyketones by means of Friedel-Craft polycondensation of bis (arylsilanes) with chlorides of aromatic dicarboxylic acids (isophthaloyl-, terephthaloyl-, 4,4{}-oxydibenzoylchloride) at 20 °C in environment of dissolvent (1,2-dichlorethane) in the presence of aluminum chloride is proposed in [331]. The polyketones have the intrinsic viscosity more than 0,37 dL/g (at 30 °C, in concentrated H_2SO_4), glass transition temperature is 120–231 °C and melting point lies within 246–367 °C. The polyketones start decomposing at a temperature of 400 °C, the temperature of 10% loss of mass is 480–530 °C.

It is possible to produce polyketones by the reaction of aromatic dicarboxylic acid and aromatic compound containing two reactive groups [332]. The reaction is catalyzed by the mix of phosphoric acid and carboxylic acid anhydride having the formula of RC(O)O(O)CR (R stands for not-substituted or substituted alkyl, in which one, several or all hydrogen atoms were replaced by functional groups and each R has the Gamet constant $\sigma_m \geq 0,2$). The pressed articles can be created from synthesized polyketones.

The aromatic polyketones can be synthesized [333] when conducting the Friedel-Craft polymerization with acylation by means of the interaction of 2,2{}bis(arylphenoxy)bisphenyls with chlorides of arylenebicarbonic acids in the presence of $AlCl_3$. The polyketones obtained using the most efficient 2,2{}bis(4-bezoylphenoxy) bisphenyl are well soluble in organic solvents and possess high heat resistance.

Also, the aromatic polyketones are produced [334] by the reaction of electrophylic substitution (in dispersion) of copolymer of aliphatic vinyl compound (1-acosen) with N-vinylpirrolydone, at ratio of their links close to equimolar.

The other method of synthesizing aromatic polyketones includes the interaction of monomer of formula $[HC(CN)(NR_2)]Ar$ with 4,4'-difluorobisphenylsulfone in dimethylsulfoxide, dimethylformamide or N-methylpirrolydone at 78–250 °C in the presence of the base [335]. The soluble polyaminonitrile of formula $[C(CN)(NR_2)ArC(CN)(NR_2)C_6H_4SO_2C_6H_4]_n$ is thus produced (Ar denotes $m-C_6H_4$, NR_2 stands for group-$NCN_2CH_2OCH_2CH_2$). When acid hydrolysis of polyaminonitrile is carried out the polysulfoneketone of formula $[COArCOC_6H_4SO_2C_6H_4]_n$ forms with glassing temperature of 192 °C, melting point 257 °C and when 10% mass loss happens at 478 °C. The acid hydrolysis can be performed in the presence of n-toluenesulfo acid, trifluoroacetic acid, and mineral acids.

The synthesis of aromatic polyketone particles has been carried out [336] by means of precipitation polycondensation and is carried out at very the low concentration of monomer [0.05 mol/L]. The polyketones are produced from bisphenoxybenzophenone (0.005 mol) or isophthaloylchloride (0.005 mol) in 100 milliliters of 1,2-dichloroethane. Some of obtained particles have highly organized the needle-shaped structure (the whisker crystals). The use of isophthaloyl instead of terephthaloyl at the same low concentration of monomer results in formation of additionally globular particles, the binders of strip structures gives rise. The average size of needle-shaped particles is 1–5 mm in width and 150–250 mm in length.

The synthesis of aromatic high-molecular polyketones by the low-temperature solid-state polycondensation of 4,4'-bisphenoxybenzophenone and isophthaloylchloride in the presence of $AlCl_3$ in 1,2-dichlorethane is possible [337]. Obtained polymers are thermoplasts with glass transition temperature 160 °C and melting point 382 °C.

The polyesterketones are synthesized [338] from dichlorbenzophenone and Na_2CO_3 in the presence of catalyst SiO_2-Cu-salt. The polymer has the negligible number of branched structures and differs from polyesterketones synthesized on the basis of 4-hydroxy-4'-flourobenzophenone on physical properties (its pressed samples have higher crystallinity and orientation).

There exist reports on syntheses of fully aromatic polyketones without single ester bonds [339–349].

The fully aromatic polyketones without ether bonds were produced [339] on the basis of polyaminonitrile, which was synthesized from anions of bis(aminonitrile) and 4,4'{}-difluorobenzophenone using the sodium hydride in mild conditions. The acid hydrolysis of synthesized polyaminonitrile avails one to obtain corresponding polyketone with high thermal properties and tolerance for organic solvents.

They are soluble only in strong acids such as concentrated H_2SO_4. The polyketones have glassing temperature 177–198 °C. Their melting points and temperature of the beginning of decomposition are respectively 386–500 °C and 493–514 °C [340].

The aromatic polyketones without ester bonds can also be produced by the polymerization of bis(chlorbenzoyl)dimethoxybisphenyls in the presence of nickel compounds [341]. The polymers have the high molecular mass, the amorphous structure, the glass transition temperature 192 °C and 218 °C and form abiding flexible films.

The known aromatic polyketones (the most of them) dissolve in strong acids, or in trifluoroacetic acid mixed with methylenechloride, or in trifluoroacetic acid mixed with chloroform.

It is known that the presence of bulk lateral groups essentially improves the solubility of polyketones, and also improves their thermo-stability. In connection to this, dichloranhydride of 3,3-bis-(4'-carboxyphenyl)phtalide (instead of chloranhydride) was used as initial material for condensation with aromatic hydrocarbons [350]. According to the Friedel-Crafts reaction of electrophylic substitution in variant of low-temperature precipitation polycondensation the high-molecular polyarylenepthalidesterketones have been synthesized. Obtained polymers have greater values of intrinsic viscosity 1.15–1.55 dL/g (in tetrachloroethane). The softening temperatures lies within 172–310 °C, the temperature of the beginning of decomposition is 460 °C. Polyarylenepthalidesterketones dissolve in wide range of organic solvents, form colorless, transparent, abiding (s = 85–120 MPA) and elastic (ε = 80–300%) films when formed from the melt.

The soluble polyketones–polyarylemethylketones–can be produced [351] by condensation polymerization of 1,4-dihalogenarenes and 1,4-diacetylbenzols in the presence of catalytical palladium complexes, base and phosphoric ligands. The high yield of polymer is seen when using the tetrahydrofurane, o-dichlorbenzole and bisphenyl ester as solvents. The synthesized polymers dissolve in tetrahydrofurane, dichloromethane and hexane, has the decomposition point 357 °C (in nitrogen). It has luminescent properties: emits the green light (490–507 nm) after light irradiation with wavelength of 380 nm.

The Refs. [352–355] report on syntheses of carding aromatic polyketones.

The ridge-like polyarylesterketones have been synthesized by means of one-stage polycondensation in solution of bis(4-nitrophenyl)ketone with phenolphthalein, o-cresolphthaleine, 2,5,2'-5'-tetramethylphenolphthaleine and timolphthaleine [352]. Authors have shown that the free volume within the macromolecule depends on position, type and number of alkyl substitutes.

The homo-and copolyarylenesterketones of various chemical constitution (predominantly, carding ones) have been produced by the reaction of nucleophylic substitution of aromatic activated dihaloid compound [353], and also the "model" homopoyarylenesterketones on the basis of bisphenol, able to crystallize, created. The tendency to crystallization is provided by the combination of fragments of carding bisphenols with segments of hydroquinone (especially, of 4,4'-dihydroxydiphenyl) and increases with elongation of difluoro-derivative (the oligomer homologues of benzophenone), and also in the presence of bisphenyl structure in that fragment. Owing to the presence of carding group, the glass transition temperature of copolymer reaches 250 °C, and the melting point–300–350 °C.

The crystallizing carding polyarylenesterketones, in difference from amorphous ones, dissolve in organic solvents very badly, they are well soluble in concentrated sulfuric acid at room temperature and when heating to boiling in m-cresol (precipitate on cooling) [354, 355].

There are contributions [356, 357] devoted to the methods of synthesis of aromatic polyketones on the basis of diarylidenecycloalkanes.

The polyketones are produced [356] by the reaction of 2,7-dibenzylidene-cyclopentanone (I) and dibenzylideneacetone (II) with dichlorides of different acids (isophthalic, 3,3{}-azodibenzoic and others) in dry chloromethane in the presence of $AlCl_3$. The "model" compounds from I and II and benzoylchloride are also obtained. The synthesized polyketones have intrinsic viscosity 0,36–0,84 dL/g (25 °C, H_2SO_4). They do not dissolve in most organic solvents, dissipate in H_2SO_4. It is determined that the polyketones containing the aromatic links are more stable than those involving aliphatic and azo group. The temperature of 10% and 50% mass loss is 150–250 °C and 270–540 °C for these polyketones.

The polyketones, possessing intrinsic viscosity 0.76–1.18 dL/g and badly dissolving in organic solvents, can be produced by Friedel-Crafts polycondensation of diarylidenecyclopentanone or diarylidenecyclohexanone, chlorides of aromatic or aliphatic diacids, or azodibenzoylchlorides. The temperature of 10% mass loss is 190–300 °C. The in polyketones have the absorption band at wavelength 240–350 nm in ultraviolet spectra (visible range) [357].

The metal-containing polyketones, which do not dissolve in most of organic solvents and easily dissipate in proton dissolvents, have been produced by the reaction of 2,6[bis(2-ferrocenyl)methylene]cyclohexane with chlorides of dicarboxylic acids [358]. The intrinsic viscosity of metal-containing polyketones is 0.29–0.52 dL/g.

The Refs. [359, 360] report on different syntheses of isomeric aromatic polyketones. Three isomeric aromatic polyketones, containing units of

2-trifluoromethyl-and 2,2{}-dimetoxybisphenylene were synthesized in Ref. [326] by means of direct electrophylic aromatic acylated polycondensation of monomers. Two isomers of polyketone of structure "head-to tail" and "head-to head" contain the links of 2-trifrluoromethyl-4,4'{}-bisphenylene and 2,2{}-dimethoxy-5,5{}-bisphenylene.

There exist several reports on syntheses of polyarylketones, containing bisphthalasinone and methylene [361, 362], naphthalene [363–367] links; containing sulfonic groups [368], carboxyl group in side chain [369], fluorine [370–372] and on the basis of carbon monoxide and styrol or n-sthylstirol [373]. It is shown that methylene and bisphthalasinone links in main chain of polyketones are responsible for its good solubility in m-cresol, chloroform. The links of bisphthalasinone improve thermal property of polymer.

The synthesis of temperature-resistant polyketone has been carried out in Ref. [362] by means of polycondensation of 4-(3-chlor-4-oxyphenyl)-2,3-phthaloasine-1-one with 4,4'-difluorobisphenylketone.

The polymer is soluble in chloroform, N-methylpirrolydone, nitrobenzene and tetrachloroethane, its glass transition temperature is 267 °C.

The aromatic polyketones, containing 1,4-naphthalene links were produced in Ref. [363] by the reaction of nucleophylic substitution of 1-chlor-4-(4'-chlorbenzoyl)naphthalene with (i) 1,4-hydroquinone, (ii) 4,4'-iso-propylidenediphenol, (iii) phenolphthalein, (iv) 4-(4'-hydroxyphenyl) (2H)-phthalasine-1-one, respectively. All polymers are amorphous and dissipate in some organic solvents. The polymers have good thermo-stability and the high glassing temperatures.

Fluorine containing polyarylketone was synthesized on the basis of 2,3,4,5,6-pentaflourbenzoylbisphenyl esters in Ref. [370]. The polymers possesses good mechanical and dielectric properties, has impact strength, solubility and tolerance for thermooxidative destruction.

The just-produced polyketones are stabilized by treating them with acetic acid solution of inorganic phosphate [371].

The simple fluorine containing polyarylesterketones was obtained in Ref. [372] on the basis of bisphenol AF and 4,4'-difluorobenzophenone. The polymer has the glass transition temperature 163 °C and temperature of 5% mass loss 515 °C, dielectric constant 1.69 at 1 MHz, is soluble well in organic solvents (tetrahydrofurane, dimethylacetamide and others).

The Refs. [374, 375] are devoted to the synthesis of polyesterketones with lateral methyl groups.

The polyarylenesterketones have been produced by the reaction of nucleophylic substitution of 4,4{}-difluorobenzophenone with hydroquinone [374]. The synthesis is held in sulfolane in the presence of anhydrous the K_2CO_3. The increasing content of methylhydroquinone links in polymers leads to increasing of glass transition temperature and lowering of crystallinity degree, melting temperature and activation energy.

The methyl-substituted polyarylesterketones have been produced in Ref. [375] be means of electrophylic polymerization of 4,4{}bis-(0-methylphenoxy)bisphenylketone or 1,4-bis(4-(methylphenoxy)benzoyl)benzene with terephthaloyl or isophthaloyl chloride in 1,2-dichloroethane in the presence of dimethylformamides and $AlCl_3$. The polymers have glassing temperature and melting point 150–170 °C and 175–254 °C.

The simple aromatic polyesterketones, which are of interest as constructional plastics and film materials, capable of operating within the long time at 200 °C, have been synthesized [376–381] by means of polynitrosubstitution reaction of 1,1-dichlor-2, 2-di(4-nitrophenyl)-ethylene and 4,4'-dinitrobenzophenone with aromatic bisphenols.

The simple esterketone oligomer can be synthesized [378] also by polycondensation of aromatic diol with halogen-containing benzophenone at 150–250 °C in organic solvent in the presence of alkali metal compound as catalyst and water.

The particles of polyesterketone have a diameter ≤ 50 micrometers, intrinsic viscosity 0.5–2 dL/g (35 °C, the ratio n-chlorphenol: phenol is 90:10).

The simple polyesterketone containing lateral side cyano-groups, possessing the glass transition temperature in range 161–179 °C, has been produced [379] by low-temperature polycondensation of mix 2,6-phenoxybenzonitrile and 4,4'-bisphenoxybenzophenone with terephthaloylchloride in 1,2-dichloroethane.

The simple polyarylesterketones, containing the carboxyl pendent groups [380] and flexible segments of oxyethylene [381] have been synthesized either.

Russian and foreign scientists work on synthesis of copolymers [382–393] and block copolymers [394–399] of aromatic polyketones.

So, the high-molecular polyketones, having amorphous structure and high values of glassing temperature, have been produced by copolymerizing through the mechanism of aromatic combination of 5,5-bis(4-chlorbemzoyl)-2,2-dimethoxybisphenyl and 5,5-bis(3-chlorbenzoyl)-2,2-dimetoxybishenyl with help of nickel complexes [382].

Terpolymers on the basis of 4,4'-bisphenoxybisphenylsulfone, 4,4'-bisphenoxybenzophenone and terephthaloylchloride were produced in [383] by low-temperature polycondensation. The reaction was lead in solution of 1,2-dichloroethane in the presence of $AlCl_3$ and N-methyl-2-pyrrolidone. It has been found that with increasing content of links of 4,4'-bisphenoxybisphenylsulfone in copolymer their glassing and dissipation temperatures increase, but melting point and temperature of crystallization decrease.

The new copolyesterketones have been produced in [384] also from 4,4'-difluorobenzophenone, 2,2{},3,3{},6,6{}-hexaphenyl-4,4'-bisphenyl-1,1{}-diol and hydroquinone by the copolycondensation in solution (sulfolane being the dissolvent) in the presence of bases (Na$_2$CO$_3$, K$_2$CO$_3$). The synthesized copolyesterketones possess solubility, high thermal stability; have the good breaking strength and good gas-separating ability in relation to CO$_2$/N$_2$ and O$_2$/N$_2$.

The random copolymers of polyarylesterketones were produced by means of nucleophylic substitution [385] the basis of bisphenol A and carding bisphenols (in particular, phenolphthalein).

The simple copolyesterketones (copolyesterketonearylates) are synthesized [386] and the method for production of copolymers of polyarylates and polycarbonate with polyesterketones is patented [387]. The reaction is held in dipolar aprotone dissolvent in the presence of interface catalyst (hexaalkylguanidinehalogenide).

The statistical copolymers of polyarylesterketones, involving naphthalene cycle in the main chain, can be produced [388] by low-temperature polycondensation of bisphenyloxide, 4,4'-{}-bis(β-naphtoxy)benzophenone with chloranhydrides of aromatic bicarbonic acids–terephthaloylchloride and isophthaloylchloride (I) in the presence of catalytical system AlCl$_3$/N–methylpirrolydone/ClCH$_2$CH$_2$Cl (copolymers are characterized by improved thermo-and chemical stability), and also by the reaction of hydroquinone with 1,4-bis(4,4'-flourobenzoyl)naphthalene (II) in the presence of sodium and potassium carbonates in bisphenylsulfone [389].

The glass transition temperature of polyarylesterketones is going up, and melting point and temperature of the beginning of the destructions are down with increasing concentration of links of 1,4-naphthalene in main chain of copolymers.

The polyarylesterketone copolymers, containing lateral cyano-groups, can be synthesized [390] on the basis of bisphenyl oxide, 2,6-bisphenoxybenzonitrile (I) and terephthaloylchloride in the presence of AlCl$_3$, employing 1,2-dichloroethane as the dissolvent, N-mthyl-2-pyrrolidon as the Lewis base. With increasing concentration of links of I the crystallinity degree and the melting point of copolymers decrease while the glass transition temperature increase. The temperature of 5%of copolymers mass loss is more than 514 °C (N$_2$).

Copolymers, containing 30–40 wt.% of links of I, possess higher therm0-stability (350 ± 10 °C), good tolerance for action of alkalis, bases and organic solvents.

The copolymers of polyarylesterketones, containing lateral methyl groups, can be produced by low-temperature polycondensation of 2,2{}-dimethyl-4,4{}-bisphenoxy bisphenyleneketone (I) or 1,4-[4-(2-methylphenoxy) benzoyl] benzene (II) and bisphenylester, terephthaloylchloride [391]. The synthesis is held in 1,2-dichloroethane in the presence of dimethylformamides and AlCl$_3$ as catalyst. With increasing concentration of links of I or II in copolymers the glass transition temperature increases, and the melting point and the crystallinity degree decreases.

The copolymers of polyarylketones, containing units of naphthalenesulfonic acid in lateral links, are use in manufacturing of proton-exchange membranes [392].

The polymeric system, suitable as proton-exchanged membrane in fuel cells, was developed in Ref. [393] on the basis of polyarylenesterketones.

The polyarylesterketones are firstly treated with metasulfonic acid within 12 h at 45 °C up to sulfur content of 1,2%, and then sulfonize with oleum at 45 °C up to sulfur content 5% (the degree of sulfonation 51%).

The high-molecular block-copolyarylenesterketones have been synthesized on the basis of 4,4'-difluorobenzophenone and number of bisphenols by means of reaction of nucleophylic substitution of activated arylhalogenide in dimethylacetamide (in the presence of potassium carbonate) [394, 395]. It has been identified the cutback of molecular weight of polymer when using bisphenolate of 4,4'-(isopropylidene)bisphenol.

The block-copolyesterketones are synthesized [396] on the basis of di-chloranhydride-1,1'-dichlor-2,2'-bis(n-carboxyphenyl)ethylene and various dioxy-compounds with complex of worthy properties. It is possible to create, on the basis of obtained polymers, the constructive and film materials possessing lower combustibility and high insulating properties. Some synthesized materials are promising as modifiers of high density polyethylene.

The block copolymers, containing units of polyarylesterketones and segments of thermotropic liquid-crystal polyesters of different length, are synthesized by means of high-temperature polycondensation in solution [397, 398]; the kinetics, thermal and liquid-crystal properties of block-copolymers have been studied.

The polyarylenesterketones have been obtained [399], which contain blocks of polyarylesterketones, block of triphenylphosphite oxide and those of binding agent into the cycloaliphatic structures. The film materials on their basis are used as polymer binder in thermal control coatings.

The Refs. [400–402] informed on production of polyester-α-diketones using different initial compounds.

Non-asymmetrical polyester-α-diketones are synthesized [400] on the basis of 4-flouru-4'(n-fluorophenylglyoxalyl)benzophenone. Obtained polymers are amorphous materials with glassing temperature 162–235 °C and have the temperatures of 10% mass loss in range of 462–523 °C.

The polyester-α-diketones synthesized in Ref. [401] on the basis of 2,2-bis[4-(4-fluorophenylglyoxalyl)phenyl]hexafluoropropane are soluble polymers. They exhibit the solubility in dimethylformamide, dimethylacetamide, N-methylpirrolydone, tetrahydrofurane and chloroform. The values of glassing temperature lie in interval 182–216 °C, and temperatures of 10% mass loss are within 485–536 °C and 534–556 °C in air and argon, respectively.

The number of new polyester-α-diketones in the form of homopolymers and copolymers has been produced [402]. Polyester-α-diketones have the amorphous structure, include α-diketone, α-hydroxyketone and ester groups, they are soluble in organic solvents. The films with lengthening of 5–87%, solidity and module under tension respectively 54–83 MPa and 1.6–3 GPa are produced by casting from chloroform solution.

The quinoxalline polymers, possessing higher glassing and softening temperatures and better mechanical properties compared to initial polymers, have been produced by means of interaction of polyester-α-diketones with α-phenylenediamine at 23 °C in m-cresol. Polyquinoxalines are amorphous polymers, soluble in chlorinated, amide and phenol dissolvents with η_{lim} = 0.4–0.6 dL/g (25 °C, in N-methylpirrolydone of 0.5 g/dL).

It is reported [403] on liquid-crystal polyenamineketone and on mix of liquid-crystal thermotropic copolyester with polyketone [404]. The mix is prepared by melts mixing. It has been found that the blending agents are partially compatible; the mixes reveal two glass transition temperatures.

The transitions of polyketones into complex polyesters are possible.

So, when conducting the reaction at 65–85 °C in water, alcohol, ester, acetonitrile, the polyketone particles of size 0.01–100 micrometers transform the polyesters by the oxidation under the peroxide agents: peroxybenzoic, m-chloroperoxybenzoic, peroxyacetic, triflouruperoxyacetic, monoperoxyphthalic, monoperoxymaleine acids, combinations of H_2O_2 and urea or arsenic acid [405, 406].

For increasing of basic physical-mechanical characteristics and reprocessing, in particular of solubility, the synthesis of aromatic polyketones is lead through the stages of formation of oligomers with end functional groups accompanied by the production of block-copolyketones or through the one-stage copolycondensation initial monomers with production of copolyketones.

So, unsaturated oligeketons are produced [407, 408] by the condensation of aromatic esters (e.g., bisphenyl ester, 4,4'-bisphenoxybisphenyl ester and others) with maleic anhydride in the presence catalyst $AlCl_3$.

The oligoketones with end amino groups can be produced [409] on the basis of dichloranhydride of aromatic dicarboxylic acid, aromatic carbohydrate and telogen (N-acylanylode) according Friedel-Crafts reaction in organic dissolvent.

The oligoketones can be produced [410] by polycondensation of bisphenyloxide and 4-fluorobenzoylchloride in solution, in the presence of $AlCl_3$. It has been found that the structure of synthesized oligomer is crystalline.

The synthesis of oligoketones, containing phthaloyl links, can be implemented [411] by the Friedel-Craft reaction of acylation.

The cyclical oligomers of phenolphthalein of polyarylenestersulfoneketone can be produced [412] by cyclical depolymerization of corresponding polymers in dipolaraprotic solvent (dimethylformamide), dimethylacetate in the presence of CsF as catalyst.

The aromatic oligoesterketones can be produced by means of interaction between 4,4'-dichlorbisphenylketone and 1,1-dichlor-2,2-di(3,5-dibrom-n-oxyphenyl)ethylene in aprotic dipolar dissolvent (dimethylsulfoxide) at 140 °C in inert gas [413]. The copolyesterketones of increased thermo-stability, heat-and fire-resistance can be synthesized on the basis of obtained oligomers.

The works on production of oligoketones and synthesis of aromatic polyketones on their basis are held in Kh. M. Berbekov Kabardino-Balkarian State University.

So, availing the method of high-temperature polycondensation in environment of dimethylsulfoxide, the oligoketones are produced with edge hydroxyl groups with the degree of condensation 1–20 from bisphenols (diane or phenolphthalein) and dichlorbenzophenone. The condensation on the second stage is carried out at room temperature in 1,2-dichloroethane in the presence of HCl as acceptor and triethylamine as catalyst with introduction of the diacyldichloride of 1,1-dichlor-2,2-(n-carboxyphenyl)ethylene into the reaction. The reduced viscosity of copolyketones lies within 0.78–3.9 dL/g for polymers on the basis of diane oligoketones and 0.50–0.85 dL/g for polymers on the basis of phenolphthalein oligoketones. The mix of the tetrachloroethane and phenol in molar ratio 1:1 at 23 °C was taken as a solvent [414].

The aromatic block-copolyketones have been synthesized by means of polycondensation of oligoketones of different composition and structure with end OH-groups on the basis of 4,4'-dichlorbenzophenone and dichloranhydrides mixed with iso-and terephthalic acids in environment of 1,2-dichloro-ethane [415–418].

The synthesis is lead via the method of acceptor-catalyst polycondensation. The double excess of triethylamine in respect to oligoketones is used as acceptor-catalyst. Obtained block-copolyketones possess good solubility in chlorinated organic solvents and can be used as temperature-resistant high-strength durable constructional materials.

3.7 AROMATIC POLYESTERSULFONEKETONES

Along with widely used polymer materials of constructional assignment such as polysulfones, polyarylates, polyaryleneketones et cetera, the number of researches has recently, within the last decades, appeared which deal with production and study of properties of polyestersulfoneketones [290, 419–433]. The advantage of given polymeric materials is that these simultaneously combine properties of both polysulfones and polyaryleneketones and it allows one to exclude some or the other disadvantages of two classes of polymers. It is known, that high concentration of ketone groups results in greater T_{glass} and T_{flow} respectively, or–in better heat stability. On the other hand, greater concentration of ketone groups results in worsening of reprocessibility of polyarylenesterketones. That is why the sulfonation of the latter has been performed. The solubility increases with increasing of the degree of sulfonation. Polyestersulfoneketones gain solubility in dichloroethane, chloroform, dimethylformamide. The combination of elementary links of polysulfone with elementary links of polyesterketone also improves the fluidity of composition during extrusion.

The Refs. [424–433] report on various methods of production of polyestersulfoneketones.

The 4,4{}-bis (phenoxy) bisphenylsulfone has been synthesized by means of reaction of phenol with bis-(4-chlorphenyl)sulfone and the low-temperature polycondensation of the product has been carried out in the melt with tere-and isophthaloylchloride resulting in the formation of polyarylenestersulfonesterketoneketones. The polycondensation is conducted in 1,2-dichlorethane at presence of $AlCl_3$ and N-methylpirrolydone. Compared to polyesterketoneketones synthesized polymers have greater temperatures of glassing and decomposition and lower melting point and also possess higher thermal resistance [424].

The block-copolymers have been created by means of polycondensation of low-molecular polyesterketone containing the remnants of chloranhydride of carbonic acid and 4,4{}-bisphenoxybisphenylsulfone (I) as end groups.

The increase in concentration of component I result in greater glassing temperature and lower melting point of block-copolymers.

The block-copolymers containing 32.63–40.7% of component I have glassing temperature and melting point respectively 185–193°C and 322–346°C, durability and modulus in tension respectively 86.6–84.2 MPa and 3.1–3.4 GPa and tensile elongation 18.5–20.3%.

Block-copolymers have good thermal properties and could be easily reprocessed in melt [425].

Polyestersulfoneketones have been produced be means of electrophylic Friedel-Crafts acylation in the presence of dimethylformamides and anhydrous $AlCl_3$ in environment of 1,2-dichlorethane on the basis of simple 2-and 3-methylbisphenyl esters and 4,4{}-bis(4-chloroformylpheoxy)bisphenylsulfone. The copolymer has the molecular weight of 57,000–71,000, its temperatures of glassing and decomposition are 160.5–167.0°C and > 450°C respectively, the coke end is 52–57% (N_2). Copolyestersulfoneketones are well soluble in chloroform and polar dissolvents (dimethylformamide and others) and form transparent and elastic films [426].

Some papers of Chinese researchers are devoted to the synthesis and study of properties of copolymers of arylenestersulfones and esterketones [427–429].

The block-copolyesters based on oligoesterketones and oligoestersulfones of various degree of condensation have been produced in Ref. [430] by means of acceptor-catalyst polycondensation; the products contain simple and complex ester bonds.

Polyestersulfoneketones possessing lower glassing temperatures and excellent solubility have been synthesized in Ref. [431] by means of low-temperature polycondensation of 2,2{}, 6,6{}-tetramethylbisphenoxybisphenylsulfone, iso- and terephthaloylchloride in the presence of $AlCl_3$ and N-methylpirrolydone in environment of 1,2-dichlorethane. The increasing concentration of iso-and terephthaloylchloride results in greater glassing temperatures of statistical copolymers.

The manufacture of ultrafiltration membranes is possible on the basis of mix polysulfone-polyesterketone [432, 433]. Transparent membranes obtained by means of cast from solution have good mechanical properties in both dry and hydrated state and keep analogous mechanical properties after exposition in water (for 24 h at 80 °C). The maximal conductivity of membranes at 23 °C is 4.2×10^{-2} Sm/cm while it increases to 0.11 Sm/cm at 80 °C.

In spite of a number of investigations in the area of synthesis and characterization of polyestersulfoneketones, there are no data in literature on the polyestersulfoneketones of block-composition based on bisphenylolpropane (or phenolphthalein), dichlorbisphenylsulfone and dichlorbenzophenone, terephthaloyl-bis (*n*-oxybenzoylchloride).

Russian references totally lack any researches on polyestersulfoneketones, not speaking on the production of such polymers in our country. Accounting for this, we have studied the regularities of the synthesis and produced block-copolyestersulfoneketones of some valuable properties [423]. The main structure elements of the polymers are rigid and extremely thermo-stable phenylene groups and flexible, providing for the thermoplastic reprocessibility, ester, sulfone and isopropylidene bridges.

Synthesized, within given investigation, polyestersulfoneketones on the basis of phthalic acid; polyarylates on the basis of 3,5-dibromine-*n*-oxybenzoic acid and phthalic acids and copolyesters on the basis of terephthaloyl-bis(*n*-oxybenzoic) acid are of interest as heat-resistant and film materials which can find application in electronic, radioelectronic, Avia, automobile, chemical industries and electrotechnique as thermo-stable constructional and electroisolation materials as well as for the protection of the equipment and devices from the influence of aggressive media.

KEYWORDS

- **Monomers**
- ***N*-oxybenzoic acid**
- **Oligoesters**
- **Polycondensation**
- **Polyesters**
- **Synthesis**
- **Tere- and isophthalic acids**

REFERENCES

1. Sokolov, L. V. (1976). *The Synthesis of Polymers: Polycondensation Method*. Moscow: Himia, 332 p. [in Russian].
2. Morgan, P. U. (1970). *Polycondensation Processes of Polymers Synthesis*. Leningrad: Himia, 448 p. [in Russian].
3. Korshak, V. V. & Vinogradova, S. V. (1968). *Equilibrium Polycondensation*. Moscow: Himia, 441 p. [in Russian].
4. Korshak, V. V. & Vinogradova, S. V. (1972). *Non-Equilibrium Polycondensation*. Moscow: Nauka, 696 p. [in Russian].

5. Sokolov, L. B. (1979). *The Grounds for the Polymers Synthesis by Polycondensation Method.* Moscow: Himia, 264 p. [in Russian].
6. Korshak, V. V. & Kozyreva, N. M. (1979). *Uspehi Himii, 48(1),* 5–29.
7. (1977). *Encyclopedia of Polymers.* Moscow: Sovetskaia Enciklopedia, *3,* 126–138. [in Russian].
8. Korshak, V. V. & Vinogradova, S. V. (1964). *Polyarylates.* Moscow: Nauka, 68 p. [in Russian].
9. Askadskii, A. A. (1969). *Physico Chemistry of Polyarylates.* Moscow: Himia, 211 p. [in Russian].
10. Korshak, V. V. & Vinogradova, S. V. (1958). *Heterochain Polyesters.* Moscow: Academy of Sciences of USSR, 403 c. [in Russian].
11. Lee, G., Stoffi, D. & Neville, K. (1972). *Novel Linear Polymers.* Moscow: Himia, 280 p. [in Russian].
12. Sukhareva, L. A. (1987). *Polyester Covers: Structure and Properties.* Moscow: Himia, 192 p. [in Russian].
13. Didrusco, G. & Valvaszori, A. (1982). Prospettive nel Campo Bei Tecnopolimeri Tecnopolime Resine, *5,* 27–30.
14. Abramov, V. V., Zharkova, N. G. & Baranova, N. S. (1984). Abstracts of the All-Union Conference "Exploiting Properties of Constructional Polymer Materials." Nalchik, 5. [in Russian].
15. Tebbat Tom. (1975). Engineering Plastics: Wonder Materials of Expensive Polymer Plauthings *Eur. Chem. News, 27,* 707.
16. Stoenesou, F. A. (1981). Tehnopolimeri. *Rev. Chem, 32(8),* 735–759.
17. Nevskii, L. B., Gerasimov, V. D. & Naumov, V. S. (1984). In Abstracts of the All-Union Conference "Exploiting Properties of Constructional Polymer Materials." Nalchik, 3. [in Russian].
18. Mori, Hisao. (1975). *Jap. Plast, 26(8),* 23–29.
19. Karis, T., Siemens, R., Volksen, W. & Economy, J. (1987). Melt Processing of the Phba-homopolymer in Abstracts of the 194-th ACS National Meeting of American Chemical Society. New Orleans: Los Angeles, August 30–September 4, Washington, DC, 335–336.
20. Buller, K. (1984). *Heat and Thermostable Polymers.* Moscow: Himia, 343 p. [in Russian].
21. Crossland, B., Knight, G. & Wright, W. (1986). The Thermal Degradation of Some Polymers Based upon P-Hydroxybenzoic Acid. *Brit. Polym. J, 18(6),* 371–375.
22. George, E. & Porter, R. (1988). Depression of the Crystalnematic Phase Transition in Thermotropic Liquid Crystal Copolyesters. *J. Polym. Sci, 26(1),* 83–90.
23. Yoshimura, T. & Nakamura, M. (1986). Wholly Aromatic Polyester US Patent 4609720. Publ. 02.09.86.
24. Ueno, S., Sugimoto, H. & Haiacu, K. (1984). Method for Producing Polyarylates Japan Patent Application 59–120626. Publ. 12.07.84.
25. Ueno, S., Sugimoto, H. & Haiacu, K. (1984). Method for Producing Aromatic Polyesters. Japan Patent Application 59–207924. Publ. 26.11.84.
26. Yu Michael, C. Polyarylate Formation by Ester Interchange Reaction Using–Gamma Lactones as Diluent US Patent 4533720.
27. Higashi, F. & Mashimo, T. (1986). Direct Polycondensation of Hydroxybenzoic Acids with Thionylchloride in Pyridine. *J. Polym. Sci.: Polym. Chem. Ed, 24(7),* 177–1720.
28. Bykov, V. V., Tyuneva, G. A., Trufanov, A. N., et al. (1986). *Izvestia Vuzov. Khim. i Him. Technol, 29(12),* C. 20–22.
29. Tedzaki K Hiroaka. (1973). Method for Producing Polymers of Oxybenzoic Acid. Japan Patent 48–23677. Publ. 16.07.73.

30. Process for Preparation of Oxybenzoyl Polymer US Patent 3790528. Publ. 05.02.74.
31. Sima, Takeo, Yamasiro, Saiti, & Inada, Hiroo. (1973). Method for Producing Polyesters of
    n-oxybenzoic Acid. Japan Patent 48–37–37355. Publ. 10.11.73.
32. Sakano, Tsutomu, & Miesi, Takehiro. (1984). Polyester Fiber Japan Patent Application
    59–199815. Publ. 13.11.84.
33. Higashi, Fukuji, & Yamada, Yukiharu. Direct Polycondensation of Hydroxybenzoic Acids
    with Bisphenyl Chlorophosphate in the Presence of Esters J. Polym. Sci.: Polym. Chem.
    Ed.
34. Adxuma, Fukudzi. (1985). Method for Producing Complex Polyesters. Japan Patent Ap-
    plication 60–60133. Publ. 06.04.85.
35. Chivers, R. A., Blackwell, J. & Gutierrez, G. A. (1985). X-ray Studies of the Structure
    of HBA/HNA Copolyesters in Proceedings of the 2-nd Symposium "Div. Polym. Chem.
    Polym. Liq. Cryst." Washington (DC)–New York–London, 153–166.
36. Sugijama, H., Lewis, D. & White, J. Structural Characteristics, Rheological Properties,
    Extrusion and Melt Spinning of 60/40 Poly(Hydroxybenzoic Acidcoethylene Terephta-
    late).
37. Blackwell, J., Dutierrez, G. & Chivers, R. (1985). X-ray Studies of Thermotropic Co-
    polyesters in Proceedings of the 2nd Symposium "Div. Polym. Chem. Polym. Liq. Cryst."
    Washington (DC) New York–London, 167–181.
38. Windle, A., Viney, C. & Golombok, R. (1985). Molecular Correlation in Thermotropic
    Copolyesters Faraday Discuss. Chem. Soc, 79, 55–72.
39. Calundann Gordon, & Meet, W. Processable Thermotropic Wholly Aromatic Polyester
    Containing Polybenzoyl Units US Patent 4067852.
40. Morinaga Den, Inada Hiroo, & Kuratsudzi Takatozi. (1977). Method for Producing Ther-
    mostable Aromatic Polyesters Japan Patent Application 52–121626. Publ. 13.10.77.
41. Sugimoto Hiroaki, & Hanabata Makoto. (1983). Method for Producing Aromatic Copoly-
    esters Japan Patent Application 58–40317. Publ. 09.03.83.
42. Dicke Hans-Rudolf, & Kauth Hermann. Thermotrope Aromatische Polyester Mit Hoher
    Steifigkeit, Verfahren Zu Ihrer Herstellung Und Ihre Verwenaung Zur Herstellung Ven
    Formkorpern, Filamenten, Fasern und Folien Germany Patent Application 3427886.
43. Cottis Steve, G. Production of Thermally Stabilized Aromatic Polyesters US Patent
    4639504.
44. Matsumoto Tetsuo, Imamura Takayuki, & Kagawa Kipdzi. (1987). Polyester Fiber. Japan
    Patent Application 62–133113. Publ. 1987.
45. Ueno Ryudzo, Masada Kachuiasu, & Hamadzaki Yasuhira. (1987). Complex Polyesters.
    Japan Patent Application 62–68813. Publ. 28.03.87.
46. Tsai, Hond–Bing, Lee Chyun, & Chang, Nien-Shi. (1990). Effect of Annealing on the
    Thermal Properties of Poly (4-Hydrohybenzoate-Co-Phenylene Isophthalates) Macromol.
    Chem, 191(6), 1301–1309.
47. Paul, K. T. (1986). Fire Resistance of Synthetic Furniture. Detection Methods Fire and
    Materials, 10(1), 29–39.
48. Iosida Tamakiho, & Aoki Iosihisa. (1986). Fire Resistant Polyester Composition. Japan
    Patent Application 61–215645. Publ. 25.09.86.
49. Wang, Y., Wu, D. C., Xie, X. G. & Li, R. X. (1996). Characterization of Copoly(p-Hy-
    droxybenzoate/Bisphenol-A Terephthalate) by NMR-Spectroscopy. Polym. J, 28(10),
    896–900.
50. Yerlikaya Zekeriya, Aksoy Serpil, & Bayramli Erdal. (2001). Synthesis and Character-
    ization of Fully Aromatic Thermotropic Liquid-Crystalline Copolyesters Containing m-
    hydroxybenzoic Acid Units J. Polym. Sci. A, 39(19), 3263–3277.

51. Pazzagli Federico, Paci Massimo, Magagnini Pierluigi, Pedretti Ugo, Corno Carlo, Berto-lini Guglielmo, & Veracini Carlo, A. (2000). Effect of Polymerization Conditions on the Microstructure of a Liquid Crystalline Copolyester. *J. Appl. Polym. Sci, 77(1)*, 141–150.

52. Aromatic Liquid-Crystalline Polyester Solution Composition US Patent 6838546. International Patent Catalogue C 08 J 3/11, C 08 G 63/19, 2005.

53. Yerlikaya Zekeriya, Aksoy Serpil, & Bayramli Erdal. (2002). Synthesis and Melt Spinning of Fully Aromatic Thermotropic Liquid Crystalline Copolyesters Containing m-hydroxy-benzoic Acid Units. *J. Appl. Polym. Sci, 85(12)*, 2580–2587.

54. Wang Yu-Zhang, Cheng Xiao-Ting, & Tang Xu-Dong. (2002). Synthesis, Characterization, and Thermal Properties of Phosphorus-Containing, Wholly Aromatic Thermotropic Copolyesters. *J. Appl. Polym. Sci, 86(5)*, 1278–1284.

55. Liquid-Crystalline Polyester Production Method US Patent 7005497. International Patent Catalogue C 08 G 63/00, 2006.

56. Wang Jiu-fen, Zhang Na, & Li Cheng-Jie. (2005). Synthesis and Study of Thermotropic Liquid-Crystalline Copolyester. PABA.ABPA.TPA. *Polym. Mater. Sci. Technol, 21(1)*, 129–132.

57. Method of Producing Thermotropic Liquid Crystalline Copolyester, Thermotropic Liquid Crystalline Copolyester Composition Obtained by the Same Composition US Patent 6268419. International Patent Catalogue C 08 K 5/51. (2001).

58. Hsiue Lin-tee, Ma Chen-chi, M. & Tsai Hong-Bing. (1995). Preparation and Characterizations of Thermotropic Copolyesters of p-hydroxybenzoic Acid, Sebacic Acid, and Hydro-quinone. *J. Appl. Polym. Sci, 56(4)*, 471–476.

59. Frich Dan, Goranov Konstantin, Schneggenburger Lizabeth, & Economy James. (1996). Novel High-Temperature Aromatic Copolyester Thermosets: Synthesis, Characterization, and Physical Properties Macromolecules, *29(24)*, 7734–7739.

60. Dong Dewen, Ni Yushan, & Chi Zhenguo. (1996). Synthesis and Properties of Thermotropic Liquid-Crystalline Copolyesters Containing Bis-(4-oxyphenyl)Methanone. II. Co-polyesters from Bis-(4-oxyphenyl)Methanone, Terephthalic Acid, n-oxybenzoic Acid and Resorcene. *Acta Polym. 2*, 153–158.

61. Teoh, M. M., Liu, S. L. & Chung, T. S. (2005). Effect of Pyridazine Structure on Thin-Film Polymerization and Phase Behavior of Thermotropic Liquid Crystalline Copolyesters. *J. Polym. Sci. B, 43(16)*, 2230–2242.

62. Liquid Crystalline Polyesters having a Surprisingly Good Combination of a Low Melting Point, a High Heat Distortion Temperature, a Low Melt Viscosity, and a High Tensile Elongation US Patent 5969083. International Patent Catalogue C 08 G 63/00. 1999

63. Aromatic complex polyester US Patent 6890988. International Patent Catalogue C 08 L 5/3477. 2005.

64. Process for Producing Amorphous Anisotropic Melt-Forming Polymers Having a High Degree of Stretchability and Polymers Produced by Same US Patent 6207790. *International Patent Catalogue* C 08 G 63/00. 2001.

65. Process for Producing Amorphous Anisotropic Melt-Forming Polymers Having a High Degree of Stretchability and Polymers Produced by Same US Patent 6132884. *International Patent Catalogue* B 32 B 27/06. 2000.

66. Process for Producing Amorphous Anisotropic Melt-Forming Polymers having a High Degree of Stretchability US Patent 6222000. *International Patent Catalogue* C 08 G 63/00. 2001.

67. He Chaobin, Lu Zhihua, Zhao Lun, & Chung Tai-Shung. (2001). Synthesis and Structure of Wholly Aromatic Liquid-Crystalline Polyesters Containing Meta-and Ortholinkages. *J. Polym. Sci. A, 39(8)*, 1242–1248.

68. Choi Woon-Seop, Padias Anne Buyle, & Hall, H. K. (2000). LCP Aromatic Polyesters by Esterolysis Melt Polymerization. *J. Polym. Sci. A, 38(19)*, 3586–3595.
69. Chung Tai-Shung, & Cheng Si-Xue. (2000). Effect of Catalysts on Thin-Film Polymerization of Thermotropic Liquid Crystalline Copolyester. *J. Polym. Sci. A, 38(8)*, 1257–1269.
70. Collins, T. L. D., Davies, G. R. & Ward, I. M. (2001). The Study of Dielectric Relaxation in Ternary Wholly Aromatic Polyesters Polym. *Adv. Technol, 12(9)*, 544–551.
71. Shinn Ted-Hong, Lin Chen-Chong, & Lin David, C. (1995). Studies on Co [Poly(Ethylene Terephthalate-p-oxybenzoate)] Thermotropic Copolyester: Sequence Distribution Evaluated from TSC Measurements. *Polym, 36(2)*, 283–289.
72. Poli Giovanna, Paci Massimo, Magagnini Pierluigi, Schaffaro Roberto, & La Mantia Francesco, P. (1996). On the use of PET-LCP Copolymers as Compatibilizers for PET/LCP Blends. *Polym. Eng. and Sci, 36(9)*, 1244–1255.
73. Chen Yanming. (1998). The Study of Liquid-Crystalline Copolyesters PHB/PBT, Modified with HQ-TRA. *J. Fushun Petrol. Inst, 18(1)*, 26–29.
74. Wang Jiu-fen, Zhu Long-Xin, & Huo Hong-Xing. (2003). The Method for Producing Thermotropic Liquid-Crystalline Complex Copolyester of Polyethyleneterephthalate. *J. Funct. Polym, 16(2)*, 233–237.
75. Liu Yongjian, Jin Yi, Bu Haishan, Luise Robert, R. & Bu Jenny. (2001). Quick Crystallization of Liquid-Crystalline Copolyesters Based on Polyethyleneterephthalate. *J. Appl. Polym. Sci, 79(3)*, 497–503.
76. Flores, A., Ania, F. & Balta Calleja, F. J. (1997). Novel Aspects of Microstructure of Liquid Crystalline Copolyesters as Studied by Microhardness: Influence of Composition and Temperature. *Polym, 38(21)*, 5447–5453.
77. Hall, H. K.(Jr)., Somogyi Arpad, Bojkova Nina, Padias Anne, B. & Elandaloussi El Hadj. (2003). MALDI-TOF Analysis of all-Aromatic Polyesters/in PMSE Preprints. Papers Presented at the Meeting of the Division of Polymeric Materials Science and Engineering of the American Chemical Society (New Orleans, 2003). *Amer. Chem. Soc, 88*, 139–140.
78. Takahashi Toshisada, Shoji Hirotoshi, Tsuji Masaharu, Sakurai Kensuke, Sano Hirofumi, & Xiao Changfa. (2000). The Structure and Stretchability in Axial Direction of Fibers from Mixes of Liquid-Crystalline all-Aromatic Copolyesters with Polyethyleneterephthalate. *Fiber, 56(3)*, 135–144.
79. Watanabe Junji, Yuaing Liu, Tuchiya Hitoshi, & Takezoe Hideo. (2000). Polar Liquid Crystals Formed from Polar Rigid-Rod Polyester Based on Hydroxybenzoic Acid and Hydroxynaphthoic Acid. *Mol. Cryst. and Liq. Cryst. Sci. and Technol.* A, *346*, 9–18.
80. Juttner, G., Menning, G. & Nguyen, T. N. (2000). Elastizitatsmodul Und Schichtenmorphologie Von Spritzgegossenen LCP-Platten. *Kautsch. and Gummi. Kunstst, B. 53*. S. 408–414.
81. Bharadwaj Rishikesh, & Boyd Richard H. (1999). Chain Dynamics in the Nematic Melt of an Aromatic Liquid Crystalline Copolyester: A Molecular Dynamics Simulation Study. *J. Chem. Phys, 1(20)*, 10203–10211.
82. Wang, Y., Wu, D. C., Xie, X. G. & Li, R. X. (1996). Characterization of Copoly(p-Hydroxybenzoate/Bisphenol-A terephthalate) by NMR-Spectroscopy *Polym. J, 28(10)*, 896–900.
83. Ishaq, M., Blackwell, J. & Chvalun, S. N. (1996). Molecular Modeling of the Structure of the Copolyester Prepared from p-hydroxybenzoic Acid, Bisphenol and Terephthalic Acid. *Polym, 37(10)*, 1765–1774.
84. Cantrell, G. R., McDowell, C. C., Freeman, B. D. & Noel, C. (1999). The Influence of Annealing on Thermal Transitions in a Nematic Copolyester. *J. Polym. Sci. B, 37(6)*, 505–522.

85. Bi Shuguang, Zhang, Yi, Bu Haishan, Luise Robert, R. & Bu Jenny, Z. (1999). Thermal Transition of a Wholly Aromatic Thermotropic Liquid Crystalline Copolyester. *J. Polym. Sci. A, 37(20)*, 3763–3769.
86. Dreval, V. E., Al-Itavi, Kh. I., Kuleznev, V. N., Bondarenko, G. N. & Shklyaruk, B. F. (2004). *Vysokomol. Soed. A, 46(9)*, 1519–1526.
87. Tereshin, A. K., Vasilieva, O. V., Avdeev, N. N., Bondarenko, G. N. & Kulichihin, V. G. (2000). *Vysokomol. Soed. A-B, 42(6)*, 1009–1015.
88. Yamato Masafumi, Murohashi Ritsuko, Kimura Tsunehisa, & Ito Eiko. (1997). Dielectric β-Relaxation in Copolymer Ethyleneterephthalate-p-hydroxybenzoic Acid. *J. Polym. Sci. And Technol, 54(9)*, 544–551.
89. Carius Hans-Eckart, Schonhals Andreas, Guigner Delphine, Sterzynski Tomasz, & Brostow Witold. (1996). Dielectric and Mechanical Relaxation in the Blends of a Polymer Liquid Crystal with Polycarbonate. *Macromolecules, 29(14)*, 5017–5025.
90. Tereshin, A. K., Vasilieva, O. V., Bondarenko, G. N. & Kulichihin, V. G. (1995). *Influence of Interface Interaction on Rheological Behavior of Mixes of Polyethyleneterephthalate with Liquid-Crystalline Polyester.* In Abstracts of the III Russian Symposium on Liquid-Crystal Polymers. Chernogolovka, 124. [in Russian].
91. Dreval, V. E., Kulichihin, V. G., Frenkin, E. I. & Al-Itavi, Kh. I. (2000). *Vysokomol. Soed. A-B, 42(1)*, 64–70.
92. Kotomin, S. V. & Kulichihin, V. G. (1996). *Determination of the Flow Limit of LQ Polyesters with Help of Method of Parallel-Plate Compression.* In Abstracts of the 18-th Symposium of Rheology. Karacharovo, 61. [in Russian].
93. Zhang Guangli, Yan Fengqi, Li Yong, Wang Zhen, Pan Jingqi, & Zhang Hongzhi. (1996). The Study of Liquid-Crystalline Copolyesters of *n*-oxybenzoic Acid and Polyethylene-terephthalate. *Acta polym. Sin, 1*, 77–81.
94. Dreval, V. E., Frenkin, E. I., Al-Itavi, Kh. I. & Kotova, E. V. (1999). *Some Thermophysical Characteristics of Liquid-Crystalline Copolyester Based on Oxybenzoic Acid and Poly-ethyleneterephthalate at High Pressures.* In Abstracts of the IV-th Russian Symposium (involving international participants) "Liquid Crystal Polymers." Moscow, 62. [in Russian].
95. Brostow Witold, Faitelson Elena, A., Kamensky Mihail, G., Korkhov Vadim, P. & Rodin Yuriy, P. (1999). Orientation of a Longitudinal Polymer Liquid Crystal in a Constant Magnetic Field. *Polym, 40(6)*, 1441–1449.
96. Dreval, V. E., Hayretdinov, F. N., Kerber, M. L. & Kulichihin, V. G. (1998). *Vysokomol. Soed. A-B, 40(5)*, 853–859.
97. Al-Itavi, Kh. I., Frenkin, E. I., Kotova, E. V., Bondarenko, G. N., Shklyaruk, B. F., Kuleznev, V. N., Dreval, V. E. & Antipov, E. M. (2000). *Influence of High Pressure on Structure and Thermophysical Properties of Mixes of Polyethyleneterephthalate with Liquid-Crystalline Polymer.* In Abstracts of the 2-nd Russian Kargin Symposium Chemistry and Physics of Polymers in the Beginning of the 21 Century. Chernogolovka, Part 1. P. 1/13. [in Russian].
98. Garbarczyk, J. & Kamyszek, G. (2000). Influence of Magnetic and Electric Field on the Structure of IPP in Blends with Liquid Crystalline Polymers. In Abstracts of the 38-th Macromolecular IUPAK Symposium. Warsaw, *3*, 1195.
99. Dreval, V. E., Frenkin, E. I. & Kotova, E. D. (1996). *Dependence of the Volume Form the Temperature and Pressure for Thermotropic LQ-Polymers and their Mixes with Polypropylene.* In Abstracts of the 18-th Symposium on Rheology. Karacharovo, 45. [in Russian].
100. Al-Itavi, Kh. I., Dreval, V. E., Kuleznev, V. N., Kotova, E. V. & Frenkin, E. I. (2003). *Vysokomol. Soed, 45(4)*, 641–648.

101. Plotnikova, E. P., Kulichihin, E. P., Mihailova, I. M. & Kerber, M. L. (1996). *Rotational and Capillary Viscometry of Melts of Mixes of Traditional and Liquid-Crystalline Thermo-plasts*/in Abstracts of the 18-th Symposium on Rheology. Karacharovo, 117. [in Russian].

102. Kotomin, S. V. & Kulichihin, B. G. (1999). *Flow Limit of Melts of Liquid-Crystal Polyesters and their Mixtures.* In Abstracts of the IV Russian Symposium (involving international participants) "Liquid Crystal Polymers." Moscow, 63. [in Russian].

103. Park Dae Soon, & Kim Seong Hun. (2003). Miscibility Study on Blend of Thermotropic Liquid Crystalline Polymers and Polyester. *J. Appl. Polym. Sci, 87(11),* 1842–1851.

104. Bharadwaj Rishikesh, K. & Boyd Richard, H. (1999). Diffusion of Low-Molecular Penetrant into the Aromatic Polyesters: Modeling with Method of Molecular Dynamics. *Polymer, 40(15),* 4229–4236.

105. Luscheikin, G. A., Dreval, V. E. & Kulichihin, V. G. (1998). *Vysokomol. Soed. A-B, 40(9),* 1511–1515.

106. Shumsky, V. F. Getmanchuk, I. P., Rosovitsky, V. F. & Lipatov, Yu. S. (1996). *Rheological, ViscoElastic and Mechanical Properties of Mixes of Polymethylmethacrylate with Liquid-Crystal Copolyester Filled with Wire-like Monocrystals.* In Abstracts of the 18-th Symposium on Rheology. Karacharovo, 115. [in Russian].

107. Liu Yongjian, Jin Yi, Dai Linsen, Bu Haishan, & Luise Robert, R. (1999). Crystallization and Melting Behavior of Liquid Crystalline Copolyesters Based on Poly(ethyleneterephthalate). *J. Polym. Sci. A, 37(3),* 369–377.

108. Abdullaev, Kh. M., Tuichiev, Sh. T., Kurbanaliev, M. K. & Kulichihin, V. G. (1997). *Vysokomol. Soed. A-B, 39(6),* 1067–1070.

109. Li Xin-Gui. (1999). Structure of Liquid Crystalline Copolyesters from two Acetoxybenzoic Acids and Polyethyleneterephthalate. *J. Appl. Polym. Sci, 73(14),* 2921–2925.

110. Li Xin-Gui, & Huang Mei-Rong. (1999). High-Resolution Thermogravimetry of Liquid Crystalline Copoly(p-oxybenzoateethyleneterephthalate-m-oxybenzoate). *J. Appl. Polym. Sci, 73(14),* 2911–2919.

111. Guo Mingming, & Britain William, J. (1998). Structure and Properties of Naphthalene-Containing Polyesters. 4. New Insight into the Relationship of Transesterification and Miscibility. *Macromolecules, 31(21),* 7166–7171.

112. Li Xin-Gui, Huang Mei-Rong, Guan Gui-He, & Sun Tong. (1996). Glass Transition of Thermotropic Polymers Based upon Vanillic Acid, p-hydroxybenzoic Acid, and Poly(ethyleneterephthalate). *J. Appl. Polym. Sci, 59(1),* 1–8.

113. Additives and Modifiers Plast. Compound, *1987–1988(4),* 10, 14–16, 18, 20, 24, 26, 28, 30, 32, 34, 36, 38–40, 42–44, 46–51.

114. Sikorski, R. & Stepien, A. (1972). Nienasycone Zywice Poliestrowe Zawierajace Cherowiec Cz. *J. Studie Problemowe.–Pr. Nauk. Inst. Technol. Organicz. i Tworzyw. Sztuczn. PWr, 7,* 3–19.

115. Takase, Y., Mitchell, G. & Odajima, A. (1986). Dielectric Behavior of Rigid–Chain Thermotropic Copolyesters. *Polym. Commun, 27(3),* 76–78.

116. Volchek, B. Z., Holmuradov, N. S., Bilibin, A. Yu, & Skorohodov, S. S. (1984). *Vysokomol. Soed. A, 26(1),* 328–333.

117. Bolotnikova, L. S., Bilibin, A. Yu, Evseev, A. K., Panov, Yu. N. Skorohodov, S. S. & Frenkel, S. Ia. (1983). *Vysokomol. Soed. A, 25(10),* 2114–2120.

118. Volchek, B. Z., Holmuradov, N. S., Purkina, A. V., Bilibin, A. Yu, & Skorohodov, S. S. (1984). *Vysokomol. Soed. A, 27(1),* 80–84.

119. Andreeva, L. N., Beliaeva, E. V., Lavrenko, P. N., Okopova, O. P., Tsvetkov, V. N., Bilibin, A. Yu, & Skorohodov, S. S. (1985). *Vysokomol. Soed. A, 27(1),* 74–79.

120. Grigoriev, A. N., Andreeva, L. N., Bilibin, A. Yu, Skorohodov, S. S. & Eskin, V. E. (1984). *Vysokomol. Soed. A, 26(8),* 591–594.
121. Grigoriev, A. N., Andreeva, L. N., Matveeva, G. I., Bilibin, A. Yu, Skorohodov, S. S. & Eskin, V. E. (1985). *Vysokomol. Soed. B, 27(10),* 758–762.
122. Bolotnikova, L. S., Bilibin, A. Yu, Evseev, A. K., Ivanov, Yu. N., Piraner, O. N., Skorohodov, S. S. & Frenkel, S. Ya. (1985). *Vysokomol. Soed. A, 27(5),* 1029–1034.
123. Pashkovsky, E. E. (1986). Abstracts of the Thesis for the Scientific Degree of Candidate of Physical and Mathematical Sciences. Leningrad, 19 [in Russian].
124. Grigoriev, A. N., Andreeva, L. N., Volkov, A. Ya, Smirnova, G. S., Skorohodov, S. S. & Eskin, V. E. (1987). *Vysokomol. Soed. A, 29(6),* 1158–1161.
125. Grigoriev, A. N., Matveeva, G. I., Piraner, O. N., Lukasov, S. V. & Bilibin, A. Yu, Sidorovich, A. V. (1991). *Vysokomol. Soed. A, 33(6),* 1301–1305.
126. (1981). Liquid Crystal Order in Polymer/Ed. A. Blumshtein. Moscow,.
127. Grigoriev, A. N., Matveeva, G. I., Lukasov, S. V., Piraner, O. N., Bilibin, A. Yu, & Sidorovich, A. V. (1990). *Vysokomol. Soed. A-B, 32(5),* 394–396.
128. Bilibin, A. Yu. (1988). *Vysokomol. Soed. B, 31(3),* 163.
129. Kapralova, V. M., Zuev, V. V., Koltsov, A. I., Skorohodov, S. S. & Khachaturov, A. S. (1991). *Vysokomol. Soed. A, 33(8),* 1658–1662.
130. Helfund, E. J. (1971). Chem. Phys, *54(11),* 4651.
131. Bilibin, A. Yu, Piraner, O. N., Skorohodov, S. S., Volenchik, L. Z. & Kever, E. E. (1990). *Vysokomol. Soed. A, 32(3),* 617–623.
132. Matveeva, G. N. (1986). Abstracts of the Thesis for the Scientific Degree of Candidate of Physical and Mathematical Sciences, 17. [in Russian].
133. Volkov, A. Ya, Grigoriev, A. I., Savenkov, A. D., Lukasov, S. V., Zuev, V. V., Sidorovich, A. V. & Skorohodov, S. S. (1994). *Vysokomol. Soed. B, 36(1),* 156–159.
134. Andreeva, L. N., Bushin, S. V., Matyshin, A. I., Bezrukova, M. A., Tsvetov, V. N., Bilibin, A. Yu, & Skorohodov, S. S. (1990). *Vysokomol. Soed. A, 32(8),* 1754–1759.
135. Stepanova, A. R. (1992). Abstracts of the Thesis for the Scientific Degree of Candidate of Chemical Sciences. Sankt–Petersburg, 24 p. [in Russian].
136. He Xiao-Hua, & Wang Xia-Yu. (2002). Synthesis and Properties of Thermotropic Liquid-Crystalline Block-Copolymers Containing Links of Polyarylate and Thermotropic Liquid-Crystalline Copolyester (HTH-6) *Natur. Sci. J.* Xiangtan Univ, *23(1),* 49–52.
137. Wang Jiu-fen, Zhu-xin, & Huo Hong-xing. (2003). *J. Funct. Polym, 16(2),* 233–237.
138. Jo Byung-Wook, Chang Jin-Hae, & Jin Jung-2. (1995). Transesterifications in a Polyblend of Poly(butylene terephthalate) and a Liquid Crystalline Polyester *Polym. Eng. and Sci, 35(20),* 1615–1620.
139. Gomez, M. A., Roman, F., Marco, C., Del Pino, J. & Fatou, J. G. (1997). Relaxations in Poly(tetramethylene terephtaloyl-bis-4-oxybenzoate): Effect of Substitution in the Mesogenic Unit and in the Flexible Spacer. *Polymer, 38(21),* 5307–5311.
140. Bilibin, A. Yu, Shepelevsky, A. A., Savinova, T. E. & Skorohodov, S. S. (1982). Terephthaloyl-Bis-*n*-Oxybenzoic Acid or its Dichloranhydride as Monomer for the Synthesis of Thermotropic Liquid-Crystalline Polymers USSR Inventor Certificate 792834. International Patent Catalogue C 07 C 63/06, C 08 K 5/09.
141. Storozhuk, I. P. (1976). *Regularities of the Formation of Poly and Oligoarylenesulfonoxides and Block-Copolymers on their Base.* Thesis for the Scientific Degree of Candidate of Chemical Sciences. Moscow, 195 p. [in Russian].
142. Rigid Polysulfones Hold at 300 F. *Jron. Age,* (1965). *195(15),* 108–109.
143. High-Temperature Thermoplastics. *Chem. Eng. Progr.,* (1965). *61(5),* 144.

144. Thermoplastic Polysulfones Strength at High Temperatures. *Chem. Eng. Progr.*, (1965). *72(10)*, 108–110.
145. Polysulfones *Brit, Plast.*, 1966. *39(3)*, 132–135.
146. Lapshin, V. V.). *Plast. Massy*, 1967, *1*, 74–78.
147. Gonezy, A. A. (1979). Polysulfon-ein Hochwarmebestandiger, Transparenter Kunststof Kunststoffe, Bild. 69, *1*, S. 12–17.
148. Thornton, E. A. (1968). Polysulfone Thermoplastics for Engineering. *Plast. Eng.*
149. Moiseev, Yu. V. & Zaikov, G. E. (1979). *Chemical Stability of Polymers in Aggressive Media*. Moscow: Himia, 288 p. [in Russian].
150. Thornton, E. A. & Cloxton, H. M. (1968). Polysulfones, Properties and Processing Characteristics *Plastics, 33(364)*, 178–191.
151. Huml, J. & Doupovcova, J. (1970). Polysulfon-Nogy Druh Suntetickych Pruskuric. *Plast. Hmoty Akanc, 7(4)*, 102–106.
152. Morneau, G. A. (1970). Thermoplastic Polyarylenesulfone That Can Be Used At 500 °F. *Mod Plast, 47(1)*, 150–152, 157.
153. Storozhuk, I. P. & Valetsky, P. M. (1978). *Chemistry and Technology of High-Molecular Compounds, 2*, 127–176.
154. Benson, B. A., Bringer, R. P. & Jogel, H. A. (1967). Polymer 360, a Thermoplastic for Use at 500 °F. Presented at SPE Antes, Detroit, Michigan.
155. Jdem, A. (1967). Phenylene Thermoplastic for Use at 500 °F. *SPE Journal.*
156. Besset, H. D., Fazzari, A. M. & Staub, R. B. (1965). Plast. Technol, *11(9)*, 50.
157. Jaskot, E. S. (1966). *SPE Journal, 22*, 53.
158. Leslie, V. J. (1974). Properties Et Application Des Polysulfones. *Rev. Gen. Caontch, 51(3)*, 159–162.
159. Bringer, R. P. & Morneau, G. A. (1969). Polymer 360, a New Thermoplastic Polysulfone for Use at 500 °F Appl. *Polym. Symp, (11)*, 189–208.
160. Andree, U. (1974). Polyarilsulfon Ein Ansergewohnliecher Termoplast Kunststof Kunststoffe, *Bild. 64, (11)*, S. 684.
161. Giorgi, E. O. (1971). Termoplastico De Engenharia Ideal Para as Condicoes Brasileiras. *Rev. Guim. Ind, 40(470)*, 16–18.
162. Korshak, V. V., Storozhuk, I. P. & Mikitaev, A. K. (1976). *Polysulfones–Sulfonyl Containing polymers*. In *Polycondensation Processes and Polymers*. Nalchik, 40–78. [in Russian].
163. Two Tondh Resistant plastic Sthrive in Hot Environments Prod. Eng., (1969). *40(14)*, 112.
164. Polysulfonic Aromatici. *Mater. Plast. Ed Elast*, (1972). *38(12)*, 1043–1044.
165. Rose, J. B. (1974). *Polymer, 15(17)*, 456–465.
166. Rigby, R. B. (1979). Victrex–Polyestersulfone. *Plast. Panorama Scand, 29(11)*, 10–12.
167. Gonozy, A. A. (1979). Polysulfon-ein Hochwarmebeston Dider Transparenter Kunststoff Kunststoffe, Bild. 69, *1*, S. 12–17.
168. Un Nuovo Tecnotermoplastico in Polifenilsulfone Radel. *Mater. Plast Ed Elast.*, (1977). *2*, 83–85.
169. Polyestersulfon in Der BASF Palette. Gimmi, Asbest, Kunststoffe. Bild. (1982). *35(3)*, S. 160–161.
170. Bolotina, L. M. & Chebotarev, V. P. (2003). *Plast. Massy, (11)*, 3–7.
171. Militskova, A. M. & Artemov, S. V. (1990). *Aromatic Polysulfones, Polyester(Ester)Ketones, Polyphenylenoxides and Polysulfides of NIITEHIM: Review*. Moscow, 1–43. [in Russian].
172. High-Durable Plastics Kunststoffe, Du Hart Im Nehmen Sind Technica (Suisse). Bild. (1999). *48(25–26)*. S. 16–22. [in German]

173. Kampf Rudolf. (2006). The Method for Producing Polymers by Means of Condensation in Melt (Polyamides, Polysulfones, Polyarylates etc.) Germany Patent Application 102004034708. International Patent Catalogue C 08 P 85/00.
174. Asueva, L. A. (2010). *Aromatic Polyesters Based of Terephthaloyl-Bis-(n-oxybenzoic) Acid.* Thesis for the Scientific Degree of Candidate of Chemical Sciences. Nalchik: KBSU, 129 p. [in Russian].
175. Japan Patent Application 1256524. 1989.
176. Japan Patent Application 1315421. 1995
177. Japan Patent Application 211634. 1990.
178. Japan Patent Application 1256525. 1989.
179. Japan Patent Application 12565269. 1989.
180. Macocinschi Doina, Grigoriu Aurelia, & Filip Daniela. (2002). Aromatic Polyculfones Used for Decreasing of Combustibility. *Eur. Polym. J, 38(5),* 1025–1031.
181. US Patent 6548622. 2003.
182. Synthesis and Characterization Poly(arylenesulfone)s *J. Polym. Sci. A, 2002. 40(4),* 496–510.
183. Germany Patent Application 19926778. 2000.
184. Vologirov, A. K. & Kumysheva, Yu. A. (2003). *Vestnik KBGU. Seria Himicheskih Nauk, 5,* 86. [in Russian].
185. Mackinnon Sean, M., Bender Timothy, P. & Wang Zhi Yuan. (2000). Synthesis and Properties of Polyestersulfones *J. Polym. Sci. A, 38(1),* 9–17.
186. Khasbulatova, Z. S., Asueva, L. A. & Shustov, G. B. (2009). *Polymers on the Basis of Aromatic Oligosulfones*/in *Proceedings of the X International Conference on Chemistry and Physicochemistry of Oligomers.* Volgograd, 100. [in Russian].
187. Ilyin, V. V. & Bilibin, A. Yu. (2002). *Synthesis and Properties of Multiblock-Copolymers Consisting of Flexible and Rigid-Link Blocks*/in Materials of the 3-rd Youth school-Conference on Organic Synthesis. Sankt-Petersburg, 230–231. [in Russian].
188. Germany Patent Application № *19907605.* 2000.
189. Reuter Knud, Wollbom Ute, & Pudleiner Heinz. (2000). Transesterification as Novel Method for the Synthesis of Block-Copolymers of Simple Polyester-Sulfone/in Papers of the 38-th Macromolecular IUPAC Symposium. Warsaw, 34.
190. Zhu Shenmin, Xiao Guyu, & Yan Deyue. (2001). Synthesis of Aromatic Graft Copolymers *J. Polym. Sci. A, 39(17),* 2943–2950.
191. Wu Fangjuan, Song Caisheng, Xie Guangliang, & Liao Guihong. (2007). Synthesis and Properties of Copolymers of 4.4'-Bis-(2-methylphenoxy)Bisphenylsulfone, 1,4-Bisphenoxybenzebe and Terephthaloyl Chloride. *Acta Polym. Sin, (12),* 1192–1195.
192. Ye Su-fang, Yang Xiao-hui, Zheng Zhen, Yao Hong-xi, & Wang Ming-jun. (2006). The Synthesis and Characterization of Novel Aromatic Polysulfones Polyurethane Containing Fluorine, *40(7),* 1239–1243.
193. Ochiai Bundo, Kuwabara Kei, Nagai Daisuke, Miyagawa Toyoharu, & Endo Takeshi. (2006). Synthesis and Properties of Novel Polysulfone Bearing Exomethylene Structure Eur. *Polym. J, 42(8),* 1934–1938.
194. Bolotina L. M. & Chebotarev, V. P. (2007). The Method for Producing the Statistical Copolymers of Polyphenylenesulfidesulfones RF Patent 2311429. *International Patent Catalogue* C 08 G 75/20.
195. Kharaev, A. V., Bazheva, R. Ch, Barokova, E. B., Istepanova, O. L. & Chaika, A. A. (2007). *Fire-Resistant Aromatic Block-Copolymers Based on 1,1-Dichlor-2,2-Bis(n-oxyphenyl) Ethylele*/in Proceedings of the 3-rd Russian Scientific and Practical Conference. Nalchik, 17–21. [in Russian].

196. Saxena Akanksha, Sadhana, R., Rao, V. Lakshmana Ravindran, P. V. & Ninan K. N. (2005). Synthesis and Properties of Poly(ester nitrile sulfone) Copolymers with Pendant Methyl Groups. *J. Appl. Polym. Sci, 97,* 1987–1994.

197. Linares, A. & Acosta, J. L. (2004). Structural Characterization of Polymer Blends Based on Polysulfones *J. Appl. Polym. Sci, 92(5),* 3030–3039.

198. Ramazanov, G. A., Shahnazarov, R. Z. & Guliev, A. M. (2005). *Russian. J. Appl. Chem, 78(10),* 1725–1728.

199. Zhao Qiuxia, & Hanson James, E. (2006). Direct Synthesis of Poly(arylmethyl sulfone) Monodendrons. *Synthesis, 3,* 397–399.

200. Cozan, V. & Avram, E. (2003). Liquid-Crystalline Polysulfone Possessing Thermotropic Properties *Eur. Polym. J, 39(1),* 107–114.

201. Dass, N. N. (2000). *Indian J. Phys. A, 74(3),* 295–298.

202. Zhang Qiuyu, Xie Gang, Yan Hongxia, Xiao Jun, & Li Yurhang. (2001). The Effect of Compatibility of Polysulfone and Thermotropic Liquid-Crystalline Polymer *J. North-West. Polytechn. Univ, 19(2),* 173–176.

203. Magagnini, P. L., Paci, M., La Mantia, F. P., Surkova, I. N. & Vasnev, V. A. (1995). Morphology and Rheology of Mixes from Sulfone and Polyester Vectra–A 950 *J. Appl. Polym. Sci, 55(3),* 461–480.

204. Garcia, M., Eguiazabal, J. L. & Nuzabal, J. (2004). Morphology and Mechanical Properties of Polysulfones Modified with Liquid-Crystalline Polymer *J. Macromol. Sci. B, 43(2),* 489–505.

205. RF Patent Application 93003367/04. 1996.

206. Wang Li-Jiang, Jian Xi-Gao, Liu Yan-Jun, & Zheng Guo-Dong. (2001). Synthesis and Characterization of Polyarylestersulfoneketone from 1-Methyl-4,5-Bis(chlorbenzoyl)-Cyclohexane and 4-(4-hydroxyphenyl)-2,3-Phthalasin-1-One *J. Funct. Polym, 14(1),* C. 53–56.

207. Lei Wei, & Cai Ming-Zhong. (2004). Synthesis and Properties of Block-Copolymers of Polyesterketoneketone and 4,4{}-Bisphenoxybisphenylsulfone. *J. Appl. Chem, 21(7),* 669–672.

208. Tong Yong-Fen, Song Cai-Sheng, Wen Hong-Li, Chen Lie, & Liu Xiao-Ling. (2005). Synthesis and Properties of Copolymers of Arylestersulfones and Esteresterketones Containing Methyl Replacers. *Polym. Mater. Sci. Technol, 21(2),* 162–165.

209. Bowen W. Richard, Doneva Teodora, A. & Yin, H. B. (2000). Membranes Made from Polysulfone Mixed with Polyesteresterketone: Systematic Synthesis and Characterization. Program and Abstr. Tel Aviv, 266.

210. Zinaida S. Khasbulatova, Luisa A. Asueva, Madina A. Nasurova, Arsen M. Karayev, & Gennady B. Shustov. (2006). Polysulfonesterketones on the Oligoester Base, Their Thermo-and Chemical Resistance, 99–105.

211. Khasbulatova, Z. S., Asueva, L. A., Nasurova, M. A., Kharaev, A. M. & Temiraev, K. B. (2005). *Simple Oligoesters: Properties and Application.* In Proceedings of the 2-nd Russian Scientific and Practical Conference. Nalchik, 54–57. [in Russian].

212. Khasbulatova, Z. S., Asueva, L. A., Nasurova, M. A., Shustov, G. B., Temiraev, K. B., Kharaeva, R. A. & Asibokova, O. R. (2006). *Synthesis and Properties of Aromatic Oligoesters.* In Materials of International Conference on Organic Chemistry "Organic Chemistry from Butlerov and Belshtein till Nowadays." Sankt-Petersburg, 793–794. [in Russian].

213. Iucke, A. (1990). Polyarylketone (PAEK) Kunststoffe, Bild. *80(10),* S. 1154–1158, 1063.

214. Khirosi, I. (1983). Polyesterketone Victrex *PEEK, 31(6),* 31–36.

215. Teruo, S. (1982). Properties and Application of Special Plastics. Polyesteresterketone. Koge Dzaire, *30(9),* 32–34.

216. Hay, I. M., Kemmish, D. I., Landford, I. J. and Rae, A. J. (1984). The Structure of Crystalline PEEK. *Polym. Commun, 25(6),* 175–179.
217. Andrew, I. Lovinger, & Davis, D. D. (1984). Single Crystals of Poly (ester-ester-ketone) (PEEK). *Polym. Commun, 25(6),* 322–324.
218. Wolf, M. (1987). Anwendungstechnische Entwicklungen Bie Polyaromaten Kunststoffe, Bild. *77(6),* S. 613–616.
219. Schlusselindustrien Fur Technische Kunststoffe Plastverarbeiter. 1987. Bild. 38, *5,* S. 46–47, 50.
220. May, R. (1984). Jn. in Proceedings of the 7-th Anme. Des. Eng. Conf. Kempston, 313–318.
221. Rigby Rhymer, B. (1984). Polyesteretperketone PEEK. *Polymer News, 9,* 325–328.
222. Attwood, T. E., Dawson, P. C. & Freeman, I. L. (1979). Synthesis and Properties of Polyarylesterketones. *Amer. Chem. Soc. Polym. Prepr, 20(1),* 191–194.
223. Kricheldorf, H. R. & Bier, G. (1984). New polymer synthesis 11 Preparation of Aromatic Poly(ester ketone)s from Silylated Bisphenols. *Polymer, 25(8),* 1151–1156
224. (1986). Polyesterketone High. *Polym. Jap., 35(4),* 380
225. (1986). High Heat Resistant Film-Talpa Japan. *Plastics Age, 24(208),* 30.
226. Takao Ia. (1988). Polyestersulfones, Polyesterketones. *Koge Dzaire End Mater, 36(12),* 120–121.
227. Takao Ia. (1990). Polyesterketones. *Koge Dzaire End Mater, 38(3),* 107–116.
228. Khasbulatoba, Z. S., Kharaev, A. M., Miritaev, A. K., et al. (1992). *Plast. Massy, 3,* 3–7.
229. Hergentother, P. M. (1987). Recent Advances in High Temperature Polymers. *Polym. J, 19(1),* 73–83.
230. Method for Producing of the Aromatic Polymer in the Presence of Inert Non-Polar Aromatic Plastificator. US Patent 4110314.
231. Method for Producing Polyesters. Germany Patent Application 2731816.
232. Aromatic Simple Polyesters. GB Patent 1558671.
233. Method for Producing Polyesters. Germany Patent Application 2749645.
234. Producing of Aromatic Simple Polyesters. GB Patent 1569603.
235. Method for Producing Aromatic Polyesters. GB Patent 1563222.
236. Producing of Simple Aromatic Polyesters Containing Microscopic Inclusions of Non-Melting Compounds. US Patent 4331798.
237. Method for Producing Aromatic Polymers. Japan Patent 57–23396.
238. Wear-Resistant, Self-Lubricating Composition. Japan Patent Application 58–109554.
239. Antifriction Composition. Japan Patent Application 58–179262.
240. Thermoplastic Aromatic Polyesterketone. Japan Patent 62–146922.
241. Composition on the Basis of Aromatic Polyarylketones. Japan Patent Application 63–20358.
242. New polyarylketones. US Patent 4731429.
243. Method for Producing Crystalline Aromatic Polyesterketone. US Patent 4757126.
244. Method for Producing High-Molecular Simple Polyesters. Japan Patent Application 63–95230.
245. Aromatic Simple Esters and Method for Producing Same. Japan Patent Application 63–20328.
246. Method for Producing Aromatic Simple Polyesters. Japan Patent Application 63–20328.
247. All-Aromatic Copolyester. Japan Patent Application 63–12360.
248. All-Aromatic Copolyesters. Japan Patent Application 63–15820.

249. Colguhoun, H. M. (1984). Synthesis of polyesterketones in Trifluoromethanesulphonic Acid: Some Structure-Reactivity Relationships. *Amer. Chem. Soc. Polym. Prepr, 25(2),* 17–18.
250. Polyesterketones. Japan Patent Application 60–144329.
251. Method for Producing Polyesterketones. Japan Patent Application 61–213219.
252. New Polymers and Method for Producing same. Japan Patent Application 62–11726.
253. Method for Producing Simple Polyesterketones. Japan Patent Application 63–75032.
254. Process for Producing Aromatic Polyesterketones. US Patent 4638944.
255. Method for Producing Crystalline Aromatic Simple Polyesterketones. Japan Patent Application 62–7730.
256. Method for Producing Thermoplastic Aromatic Simple Polyesters. Japan Patent Application 62–148524.
257. Method for Producing Thermoplastic Polyesterketones. Japan Patent Application 62–148323.
258. Thermoplastic Aromatic Simple Polyesterketones and Method Producing Same. Japan Patent Application 62–151421.
259. Method for Producing Polyarylesterketones Using Catalyst on the Basis of Sodium Carbonate and Salt of Organic Acid. US Patent 4748227.
260. Thermostable Polyarylesterketones. Germany Patent Application 37008101.
261. Simple Aromatic Polyesterketones. Japan Patent Application 63–120731.
262. Impact Strength Polyarylesterketones. Japan Patent Application 63–120730.
263. Method for Producing Simple Polyarylesterketones in the Presence of Salts of Lanthanides, *Alkali and Alkali-Earth Metals.* US Patent 4774311.
264. Jovu, M. & Marinecsu, G. Rolicetoeteri. (1981). Produce de Policondensaze Ale 4,4−Dihidroxibenzofenonei Cu Compusi Bisclorometilate Aromatici *Rev. Chim, 32(12),* 1151–1158.
265. Sankaran, V. & Marvel, C. S. (1979). Polyaromatic Ester-Ketone-Sulfones Containing 1,3-Butadiene Units *J. Polymer Sci.: Polymer Chem. Ed, 17(12),* 3943–3957.
266. Method for Producing Aromatic Polyesterketones. Japan Patent Application 60–101119.
267. Method for Producing Aromatic Polyesterketones and Polythioesterketones. US Patent 4661581.
268. Method for Producing Aromatic Polyesterketones. Germany Patent Application 3416446.
269. Uncrosslinked-Linked Thermoplastic Reprocessible Polyesterketone and Method for its Production. Germany Patent Application 3416445 A.
270. Litter, M. J. & Marvel, C. S. (1986). Polyaromatic Esterketones and Polyaromatic Ester-Ketone Sulfonamides from 4-Phenoxy-Benzoyl Ester. *J. Polym. Sci.: Polym. Chem. Ed, 23(8),* 2205.
271. Method for Producing Aromatic Simple Poly(thio)Esterketone. Japan Patent Application 61–221228.
272. Method for Producing Aromatic Simple Poly(thio)Esterketone. Japan Patent Application 61–221229.
273. Method for Producing Polyarylketone Involving Treatment with the Diluents. US Patent 4665151.
274. Copolyesterketones. US Patent 4704448.
275. Method for Producing Aromatic Poly(thio)Esterketones. Japan Patent Application 62–146923.
276. Method for Producing Simple Aromatic Polythioesterketones. Japan Patent Application 62–119230.
277. Production of Aromatic Polythioesterketones. Japan Patent Application 62–241922.

278. Method for Producing Polyarylenesterketones. US Patent 4698393.
279. Method for Producing Aromatic Polymers. US Patent 4721771.
280. Method for Producing Aromatic Simple Poly(thio)Esterketones. Japan Patent Application 63–317.
281. Method for Producing Aromatic Simple Poly(thio)Esterketones. Japan Patent Application 63–316.
282. Method for Producing Polyarylesterketones. US Patent 471611.
283. Gileva, N. G., Solotuchin, M. G. & Salaskin, S. N. (1988). Synthese Von Aromatischen Polyketonen Durch Fallungspolukondensation. *Acta Polym, Bild. 39(8)*, S. 452–455.
284. Lee, I. & Marvel, S. (1983). Polyaromatic Esterketones from o, o-Disubstituted Diphenyl Esters *J. Polym. Sci.: Polym. Chem. Ed, 21(8)*, 2189–2195.
285. Method for Producing Polyarylenesterketones by Means of Electrophylic Polycondensation. Germany Patent Application 3906178.
286. Colgupoum, H. M. & Lewic, D. F. (1988). Aromatic Polyesterketones Via Superacid Catalysis/in "Spec. Polym. 88": Abstracts of the 3-rd International Conference on. *New Polymeric Materials.* Guildford, 39.
287. Colgupoum, H. M. & Lewic, D. F. (1988). Synthesis of Aromatic Polyester-Ketones in Triflouromethanesulphonic Acid. *Polym, 29(10)*, 1902.
288. Durvasula, V. R., Stuber, F. A. & Bhattacharyee, D. (1988). Synthesis of Polyphenyleneester and Thioester Ketones. *J. Polym. Sci. A, 27(2)*, 661–669.
289. Method for Producing High-Molecular Polyarylenesulfideketone. US Patent 47182122.
290. Ogawa, T., & Marvel, C. S. (1985). Polyaromatic Esterketones and Ester-Ketone-Sulfones having Various Hydrophilic Groups. *J. Polym. Sci.: Polym. Chem. Ed, 23(4)*, 1231–1241.
291. Percec, V. & Nava, H. (1988). Synthesis of Aromatic Polyesters by Scholl Reaction 1, Poly(1,1-Dinaphthyl Ester Phenyl Ketones). *J. Polym. Sci. A, 26(3)*, 783–805.
292. Mitsuree, U. & Nasaki, S. (1987). Synthesis of Aromatic Poly(ester ketones) Macromolecules, *20(11)*, 2675–2677.
293. Method for Producing Simple Polyesterketones. Japan Patent Application 61–247731.
294. Aromatic Polyesterketone and its Production. Japan Patent Application 61–143438.
295. Crystalline Polymers with Aromatic Ketone, Simple Ether and Thioether Linkages within the Main Chain and Method for Producing Same. Japan Patent Application 61–141730.
296. Producing of Crystalline Aromatic Polysulfidesterketones. Japan Patent Application 62–529.
297. Producing of Crystalline Aromatic Polyketone with Simple Ether and Sulphide Linkages. Japan Patent Application 62–530.
298. Patel, H. G., Patel, R. M. & Patel, S. R. (1987). Polyketothioesters from 4,4-Dichloro-acetyldiphenylester and their Characterization. *J. Macromol. Sci. A, 24(7)*, 835–340.
299. Method for Producing Aromatic Polyesterketones. Japan Patent Application 62–220530.
300. Method for Producing Aromatic Polyesterketones. Japan Patent Application 62–91530.
301. Aromatic Copolyketones and Method for Producing Same. Japan Patent Application 63–10627.
302. Crystalline Aromatic Polyesterketones and Method for Producing same. Japan Patent Application 61–91165.
303. Aromatic Polyesterthioesterketones and Method for Producing same. Japan Patent Application 61–283622.
304. Producing of Polyarylenoxides using Carbonates of Alkali-Earth Metals, Salts of Organic Acids and, in Some Cases, salts of Copper as catalysts. US Patent 4774314.
305. Method for Producing Polyarylenesterketones US Patent 4767837.

306. Heat-Resistant Polymer and Method for Producing same. Japan Patent Application 62–253618.
307. Heat-Resistant Polymer and Method for Producing Same. Japan Patent Application 62–253619.
308. Aromatic Polyesterketones. US Patent 4703102.
309. Producing of Aromatic Polymers. GB Patent 1569602.
310. New Polymers and Method for their Production. Japan Patent Application 61–28523.
311. Aromatic Simple Polyesterketones with Blocked end Groups and Method for Producing Same Japan Patent Application 61–285221.
312. Aromatic Simple polyesterketones and Method for their Production. Japan Patent Application 61–176627.
313. Method for Producing Crystalline Aromatic simple Polyesterketones. Japan Patent Application 62–7729.
314. Method for Producing Fuse Aromatic Polyesters. US Patent 4742149.
315. Films from Aromatic Polyesterketones Germany. Patent Application 3836169.
316. Method for Producing Polyarylenestersulfones and Polyarylenesterketones. Germany Patent Application 3836582.
317. Method for Producing Polyarylenesterketones. Germany Patent Application 3901072.
318. Polyarylesterketones. US Patent 4687833.
319. Method for Producing Oligomer Aromatic Simple Ethers. Poland Patent 117224.
320. Polymers Containing Aromatic Groupings. GB Patent 1541568.
321. Corfield, G. C. & Wheatley, G. W. (1988). The Synthesis and Properties of Blok Copolymers of Polyesteresterketone and Polydimethylsiloxane. In "Spec. Polym. 88": Abstracts of the 3-rd International Conference on New Polymeric Materials. Cambridge, 68.
322. Method for Producing Polyarylesterketones. Germany Patent Application 3700808.
323. Block-Copolymers Containing Polyarylesterketones and Methods for their Production. US Patent 4774296.
324. Poly(arylesterketones) of Improved Chain. US Patent 4767838.
325. New Block-Copolymer Polyarylesterketone-Polyesters. US Patent 4668744.
326. Simple Polyarylesterketone Block-Copolymers. US Patent 4861915.
327. Producing of Polyarylenesterketones by Means of Consecutive Oligomerization and polycondensation in Separate Reaction Zones. US Patent 4843131.
328. Khasbulatova, Z. S., Kharaev, A. M., Mikitaev, A. K., et al. (1990). *Plast Massy, (11),* 14–17.
329. Khasbulatova, Z. S. (1989). *Diversity of Methods for Synthesizing Polyesterketones/*in Abstracts of the II Regional Conference "Chemists of the Northern Caucasus–to National Economy." Grozny, 267. [in Russian].
330. Reimer Wolfgang. (1999). Polyarylesterketone (PAEK) Kunststoffe, Bild. 89, *(10),* S. 150, 152, 154.
331. Takeuchi Hasashi, Kakimoto Masa-Aki, & Imai Yoshio. (2002). Novel Method for Synthesizing Aromatic Polyketones from Bis(arylsilanes) and Chlorides of Aromatic Bicarbonic Acids *J. Polym. Sci. A, 40(16),* 2729–2735.
332. Process for Producing Polyketones. US Patent 6538098. International Patent Catalogue C 08 П 6/00, 2003.
333. Maeyata Katsuya, Tagata Yoshimasa, Nishimori Hiroki, Yamazaki Megumi, Maruyama Satoshi, & Yonezawa Noriyuki. (2004). Producing of aromatic polyketones on the Basis of 2,2{}-Diaryloxybisphenyls and Derivatives of Arylenecarbonic Acids Accompanied with Polymerization with Friedel-Krafts Acylation React. *And Funct. Polym, 61(1),* 71–79.

334. Daniels, J. A. & Stephenson, J. R. (1995). Producing of Aromatic Polyketones. GB Patent Application 2287031. International Patent Catalogue C 08 G 67/00.
335. Gibeon Harry, W. & Pandya Ashish. (1994). Method for Producing Aromatic Polyketones. US Patent 5344914. International Patent Catalogus C 08 G 69/10.
336. Zolotukhin, M. G., Baltacalleja, F. J., Rueda, D. R. & Palacios, J. M. (1997). Aromatic Polymers Produced by Precipitate Condensation *Acta Polym, 48(7),* 269–273.
337. Zhang Shanjy, Zheng Yubin, Ke Yangchuan, & Wu Zhongwen. (1996). Synthesis of Aromatic Polyesterketones by Means of Low-Temperature Polycondensation Acta Sci. *Nature. Univ. Jibimensis, 1,* 85–88.
338. Hachya Hiroshi, Fukawa Isaburo, Tanabe Tuneaki, Hematsu Nobuyuki, & Takeda Kunihiko. (1999). Chemical Structure and Physical Properties of Simple Polyesterketone Produced from 4,4'-Dichlorbenzophenone and Sodium Carbonate Trans. *Jap. Soc. Mech. Eng.* A, *65(632),* 71–77.
339. Yang Jinlian, & Gibson Harry W. (1997). Synthesis of Polyketones Involving Nucleophylic Replacement Through Carb-Anions Obtained From Bis(α-aminonitriles) Macromolecules, *30(19),* 64–73.
340. Yang Jinlian, Tyberg Christy S. & Gibson Harry, W. (1999). Synthesis of Polyketone Containing Nucleophylic Replacers Through Carb-Anions Obtained from Bis(α-aminonitriles). Aromatic polyesterketones Macromolecules, *32, (25),* 8259–8268.
341. Yonezawa Noriyuki, Ikezaki Tomohide, Nakamura Niroyuki, & Maeyama Katsuya. (2000). Successful Synthesis of all-Aromatic Polyketons by means of Polymerization with Aromatic Combination in the Presence of Nickel. *Macromolecules, 33(22),* 8125–8129.
342. (2005). Aromatic Polyesterketones US Patent 6909015. International Patent Catalogue C 07 C 65/00.
343. Toriida Masahiro, Kuroki Takashi, Abe Takaharu, Hasegawa Akira, Takamatsu Kuniyuki, Taniguchi Yoshiteru, Hara Isao, Fujiyoshi Setsuko, Nobori Tadahito, & Tamai Shoji. (2004). Patent Applicaiton 1464662. International Patent Catalogue C 08 G 65/40.
344. Richter Alexander, Schiemann Vera, Gunzel Berna, Jilg Boris, & Uhlich Wilfried. (2007). Verfahren Zur Herstellung Von Polyarylenesterketon Germany Patent Application 102006022442. International Patent Catalogue C 08 G 65/40.
345. Chen Liang, Yu Youhai, Mao Huaping, Lu Xiaofeng, Yao Lei, & Zhang Wanjin. (2005). Synthesis of a new Electroactive Poly(aryl ester ketone) Polymer, *46(8),* 2825–2829.
346. Maikhailin, Yu. A. (2007). *Polymer. Mater: Articles, Equip. Technol,* 5, C. 6–15.
347. Sheng Shouri, Kang Yigiang, Huang Zhenzhong, Chen Guohua, & Song Caisheng. (2004). Synthesis of Soluble Polychlorreplaced Polyarylesterketones Acta Polym. *Sin,* 5, 773–775.
348. Kharaev, A. M., Mikitaev, A. K. & Bazheva, R. Ch. (2007). *Halogen-Containing Polyarylenesterketones*/in Proceedings of the 3-rd Russina Scientific and Practical Conference "Novel Polymeric Composite Materials." Nalchik, 187–190. [in Russian].
349. Liu Baijun, Hu Wei, Chen Chunhai, Jiang Zhenhua, Zhang Wanjin, Wu Zhongwen, & Matsumoto Toshihik. (2004). Soluble Aromatic Poly(ester ketone)s with a Pendant 3,5-Ditrifluoromethylphenyl Group *Polymer, 45(10),* 3241–3247.
350. Gileva, N. G., Zolotukhin, N. G., Sedova, E. A., Kraikin, V. A. & Salazkin, S. N. (2000). *Synthesis of Polyarylenephthalidesterketones*/in Abstracts of the 2-nd Russian Kargin Symposium Chemistry and Physics of Polymers in the Beginning of the 21 Century. *Chernogolovka, Part 1. P. 1/83.* [in Russian].
351. Wang Dekun, Wei Peng, & Wu Zhe. (2000). Synthesis of Soluble Polyketones and Polyarylenevinylens–new Reaction of Polymerization *Macromolecules, 33(18),* 6896–6898.

352. Wang Zhonggang, Chen Tianlu, & Xu Jiping. (1995). Synthesis and Characteristics of Card Polyarylesterketones with Various Alkyl Replacers. *Acta Polym. Sin, 4*, 494–498.

353. Salazkin, S. N., Donetsky, K. I., Gorshkov, G. V., Shaposhnikova, V. V., Genin, Ya. V. & Genina, M. M. (1997). *Vysokomol. Soed. A-B, 39*, C. 1431–1437.

354. Salazkin, S. N., Donetsky, K. I., Gorshkov, G. V. & Shaposhnikova, V. V. (1996). *Doklady RAN, 348(1)*, C. 66–68.

355. Donetsky, K. I. (2000). Abstracts of the Thesis for the Scientific Degree of Candidate of Chemical Sciences. Moscow, 24 p. [in Russian].

356. Khalaf Ali, A., Aly Kamal, L. & Mohammed Ismail, A. (2002). New Method for Synthesizing Polymers *J. Macromol. Sci. A, 39, 4*, 333–350.

357. Khalaf Ali, A. & Alkskas, I. A. (2003). Method for Synthesizing Polymers. *Eur. Polym. J, 39(6)*, 1273–1279.

358. Aly Kamal, L. (2004). Synthesis of Polymers *J. Appl. Polym. Sci, 94(4)*, 1440–1448.

359. Chu, F. K. & Hawker, C. J. (1993). Different Syntheses of Isomeric Hyperbranched Polyesterketones. *Polym.* Bull, *30(3)*, 265–272.

360. Yonezawa Noriyuki, Nakamura Hiroyuki, & Maeyama Katsuya. (2002). Synthesis of all-Aromatic Polyketones having Controllable Isomeric Composition and Containing Links of 2-Trifluorometylbisphenylene and 2,2{}-Dimetoxybisphenylene. *React. And Funct. Polym, 52(1)*, 19–30.

361. Zhang Shaoyin, Jian Xigao, Xiao Shude, Wang Huiming, & Zhang Jie. (2002). Synthesis and Properties of Polyarylketone Containing Bisphthalasinone and Methylene Groupings. *Acta Polym. Sin, 6*, 842–845.

362. Chen Lianzhou, Jian Xigao, Gao Xia, & Zhang Shouhai. (1999). Synthesis and Properties of Polyesterketones Containing Links of Chlorphenylphthalasion. *Chin. J. Appl. Chem, 16(3)*, 106–108.

363. Gao Ye, & Jian Xi-gao. (2001). Synthesis and Crharacterization of Polyearyesterketones Contnaining 1,4-Naphthaline Linkages *J. Dalian Univ. Technol, 41(1)*, 56–58.

364. Wang Mingjing, Liu Cheng, Liu Zhiyong, Dong Liming, & Jian Xigao. (2007). Synthesis and Properties of Polyarylnithilesterketoneketones Containing Phthalasinon *Acta Polym. Sin, 9*, 833–837.

365. Zhang Yun-He, Wang Dong, Niu Ya-Ming, Wang Gui-Bin, & Jiang Zhen-Hua. (2005). Synthesis and Properties of Fluor-Containing Polyarylesterketones with Links of 1,4-Naphthylene. *Chem. J. Chin. Univ, 26(7)*, 1378–1380.

366. Kim Woo-Sik, & Kim Sang-Youl. (1997). Synthesis and Properties of Polyesters Containing Naphthalenetetracarboxylic Imide. *Macromol. Symp, (118)*, 99–102.

367. Cao Hui, Ben Teng, Wang Xing, Liu Na, LiuXin-Cai, Zhao Xiao-Gang, Zhang Wan-Jin, & Wei Yen. (2004). Synthesis and Properties of Chiral Polyarylesterketones Containing Links of 1,1{}-Bis-2-Naphtyl. *Chem. J. Chin. Univ, 25(10)*, 1972–1974.

368. Wang Feng, Chen Tianlu, Xu Jiping, Lui Tianxi, Jiang Hongyan, Qi Yinhua, Liu Shengzhou, & Li Xinyu. (2006). Synthesis and Characterization of Poly(arylene ester ketone) (co)Polymers Containing Sulfonate Groups. *Polymer, 47(11)*, 4148–4153.

369. Cheng Cai-Xia, Liu-Ling, & Song Cai-Sheng. (2002). Synthesis and Properties of Aromatic Polyesterketoneketone Containing Carboxylic Group within the Lateral. *Chain J. Jiangxi Norm. Univ. Natur. Sci. 26(1)*, 60–63.

370. 2,3,4,5,6-Pentafluorobenzoylbisphenylene Ethers and Fluor-Containing Polymers of Arylesterketones US Patent 6172181. International Patetn Catalogue C 08 П 73/24, 2001.

371. Ash, C. E. (1995). Process for Producing Stabilized Polyketones US Patent 5432220. International Patent Catalogue C 08 F 6/00.

372. Jiang Zhen-yu, Huang Hai-Rong, & Chen Jian-Ding. (2007). Synthesis and Properties of Polyarylesterketone and Polyarylestersulfone Containing Links of Hexafluoroizopropy-lydene J. E. China Univ. Sci. and Technol. Nat. Sci., 33(3), 345–349.
373. Xu Yongshen, Gao Weiguo, Li Hongbing, & Guo Jintang. (2005). Synthesis and Proper-ties of Aromatic Polyketones Based on CO and Stirol or n-ethylstirol J. Chem. Ind. and Eng. (China)., 56(5), 861–864.
374. Rao, V. L., Sabeena, P. U., Saxena Akanksha, Gopalakrishnan, C., Krishnan, K., Ravin-dran, P. V. & Ninan, K. N. (2004). Synthesis and Properties of Poly(aryl ester ester ketone) Copolymers with Pendant Methyl Groups Eur. Polym. J, 40(11), 2645–2651.
375. Tong Yong-Fen, Song Cai-Sheng, Chen Lie, Wen Hong-Li, & Liu Xiao-Ling. (2004). Synthesis and Properties of Methyl-Replaced Polyarylesterketone. Chin. J. Appl. Chem, 21(10), 993–996.
376. Koumykov, R. M., Vologirov, A. K., Ittiev, A. B. & Rusanov, A. L. (2005). Simple Aro-matic Polyesters and Polyesterketones Based on Dinitro-Derivatives of Clroral/in "Novel Polymeric Composite Materials": Proceedings of the 2-nd Russian Research-Practical Conference. Nalchik, 225–228. [in Russian].
377. Koumykov, R. M., Bulycheva, E. G., Ittiev, A. B., Mikitaev, A. K. & Rusanov, A. L. (2008). Plast. Massy, 3, С. 22–24.
378. Polyester Ketone and Method of Producing the Same US Patent 7217780. International Patent Catalogue C 08 G 14/04. 2006.
379. Li Jianying, Yu Yikai, Cai Mingzhong, & Song Caisheng. (2006). Synthesis and Properties of Simple Polyesterketonesterketone Containing Lateral Cyanogroups. Petrochem. Tech-nol, 35(12), 1179–1183.
380. Liu Dan, & Wang Zhonggang. (2008). Novel Polyaryletherketones Bearing Pendant Car-boxyl Groups and their Rare Earth Complexes. Part I. Synthesis and Characterization. Polymer, 49(23), 4960–4967.
381. Jeon In-Yup, Tan Loon-Seng, & Baek Jong-Beom. (2007). Synthesis of Linear and Hy-perbranched Poly(esterketone)s Containing Flexible Oxyethylene Spacers. Polym. Sci. A, 45(22), 5112–5122.
382. Maeyama Katsuya, Sekimura Satoshi, Takano Masaomi, & Yonezawa Noriyuki. (2004). Synthesis of Copolymers of Aromatic Polyketones React. And Funct. Polym, 58(2), 111–115.
383. Li, Wei, Cai Ming-Zhong, & Song Cai-Sheng. (2002). Synthesis of Ternary Copolymers from 4,4'-Bisphenoxybisphenylsulfone, 4,4'-Bisphenoxybenzophenone and Terephtha-loylchloride. Chin. J. Appl. Chem, 19(7), 653–656.
384. Gao Yan, Dai Ying, Jian Xigao, Peng Shiming, Xue Junmin, & Liu Shengjun. (2000). Syn-thesis and Characterization of Copolyesterketones Produced from Hexaphenyl-Replaced Bisphenylbisphenol and Hydroquinone. Acta Polym. Sin, 3, 271–274.
385. Sharapov, D. S. (2006). Abstracts of the Thesis for the Scientific Degree of Candidate of Chemical Sciences. Moscow, 25 p. [in Russian].
386. Kharaeva, R. A. & Ashibokova, O. R. (2005). Synthesis and Some Properties of Copolyes-terketones/in Proceedings of Young Scientists. Nalchik, KBSU, 138–141. [in Russian].
387. Method for Preparing Polyester Copolymers with Polycarbonates and Polyarylates. US Patent 6815483. International Patent Catalogue C 08 L 67/00. 2004.
388. Liu Xiao-Ling, Xu Hai-Yun, & Cai Ming-Zhong. (2001). Synthesis and Properties of Statistical Copolymers of Polyesterketoneketone and Polyesterketoneesterketoneketone Containing Naphthalene Cycle Wothon the Main Chain J. Jiangxi Norm. Univ. Natur. Sci. Ed, 25(4), 292–294.

389. Synthesis and Properties of Poly(aryl ester ketone) Copolymers Containing 1,4-Naphthalene Moieties. (2004). *J. Macromol. Sci. A, 41(10)*, 1095–1103.

390. Yu Yikai, Xiao Fen, & Cai Mingzhong. (2007). Synthesis and Properties of Poly(arylesterketone ketone)/poly(aryl ester ester ketone ketone) Copolymers with Pendant Cyano Groups. *J. Appl. Polym. Sci, 104(6)*, 3601–3606.

391. Tong Yong-Fen, Song Cai-sheng, Chen Lie, Wen Hong-li, & Liu Xiao-ling. (2005). Synthesis and Properties of Copolymers of Polyarylesterketone Containing Lateral Methyl Groupings. *Polym Mater. Sci. Technol. Eng, 21(4)*, 70–72, 76.

392. Gao Yan, Robertson Gilles, P., Guiver Michael, D., Mikhailenko Serguei D., Li Xiang, & Kaliaguine Serge. (2004). Synthesis of Copolymers of Polyaryleneesteresterketoneketons Containing Links of Naphthalene Sulfonic Acid within the Lateral Links, and their user at Manufacturing Proton-Exchange Membranes. *Macromolecules, 37(18)*, 6748–6754.

393. Mohwald Helmut, Fischer Andreas, Frambach Klaus, Hennig Ingolf, & Thate Sven. (2004). Verfahren zur Herstellung Eines Zum Protonenaustausch Befahigter Polymersystems Auf Der Basis Von Polyarylesterketonen Germany Patent Application 10309135. International Patent Catalogue C 08 G 8/28.

394. Shaposhnikova, V. V., Sharapov, D. S., Kaibova, I. A., Gorlov, V. V., Salazkin, S. N., Dubrovina, L. V., Bragina, T. P., Kazantseva, V. V., Bychko, K. A., Askadsky, A. A., Tkachenko, A. S., Nikiforova, G. G., Petrovskii, P. V. & Peregudov, A. S. (2007). *Vysokomol. Soed, 49(10)*, 1757–1765.

395. Shaposhnikova, V. V., Salazkin, S. N., Matedova, I. A. & Petrovskii, P. V. Polyarylenetherketones. Investigation of Approaches to Synthesis of Amorphous Blockpolymers in Abstracts of the 4-th International Symposium Molecular Order and Mobility in Polymer Systems.–St. Petersburg.–. 121.

396. Bedanokov, A. Yu. (1999). Abstracts of the Thesis for the Scientific Degree of Candidate of Chemical Sciences. Nalchik, 19 p. [in Russian].

397. Yang Yan-Hua, Dai Xiao-Hui, Zhou Bing, Ma Rong-Tang, & Jiang Zhen-Huang. (2005). Synthesis and Characterization of Block Copolymers Containing Poly(aryl ester ketone) and Liquid Crystalline Polyester Segments. *Chem. J. Chin. Univ, 26(3)*, 589–591.

398. Zhang Yun-He, Liu Qin-Hua, Niu Ya-Ming, Zhang Shu-Ling, Wang Dong, & Jiang Zhen-Hua. (2005). Properties and Crystallization Kinetics of Poly(ester ester ketone)-Co-Poly(ester ester ketone ketone) Block Copolymers. *J. Appl. Polym. Sci, 97(4)*, 1652–1658.

399. Polyarylenesterketone Phosphine Oxide Compositions Incorporation Cycloaliphatic Units for Use as Polymeric Binders in Thermal Control Coatings and Method for Synthesizing Same US Patent 7208551. International Patent Catalogue C 08 L 45/00–216.

400. Keshtov, M. L., Rusanov, A. L., Keshtova, S. V., Pterovskii, P. V. & Sarkisyan, G. B. (2001). *Vysokomol. Soed. A, 43(12)*, 2059–2070.

401. Keshtov, M. L., Rusanov, A. L., Keshtova, S. V., Schegolihin, A. N. & Petrovskii, P. V. (2001). *Vysokomol. Soed. A, 43(12)*, 2071–2080.

402. Brandukova, Natalya, E. & Vygodskii Yakov, S. (1995). Novel Poly α-Diketones and Copolymers on their Base *J. Macromol. Sci. A, 32*, 941–950.

403. Yandrasits, M. A., Zhang, A. Q., Bruno, K., Yoon, Y., Sridhar, K., Chuang, Y. W., Harris, F. W. & Cheng, S. Z. D. (1994). Liquid-Crystal Polyenamineketones Produced Via Hydrogen Bonds. *Polym. Int, 33(1)*, 71–77.

404. Mi Yongli, Zheng Sixun, Chan Chi-Ming, & Guo Qipeng. (1998). Mixes of Phenolphthalenin with Thermotropic Liquid-Crystal Copolyester. *J. Appl. Polym. Sci, 69(10)*, 1923–1931.

405. Arjunan Palanisamy. (1995). Production of Polyesters from Polyketones US Patent 5466780. International Patent Catalogue C 08 F 8/06, C 08 K 5/06.

406. Arjunan Palanisamy. (1996). Process of Transformation of Polyketones into the Complex Polyesters US Patent 55506312. International Patent Catalogue C 08 F 20/00.

407. Matyushov, V. F. & Golovan' S. V. (2003). Method for Producing Non-Saturated Oligoarylesterketones RF Patent 2201942. International Patent Catalogue C 08 G 61/12.

408. Matyushov, V. F. & Golovan, S. V. (2003). Method for Producing Non-Saturated Oligoarylesterketones RF Patent Application 2001109440/04. International Patent Catalogue C 08 G 61/12.

409. Matyushov, V. F., Golovan, S. V. & Malisheva, T. L. (2000). Method for Producing Oligoarylesterketones with End Amino-Groups Ukraine Patent 28015. International Patent Catalogue C 08 G 8/02.

410. Zhaobin Qiu, Zhishen Mo, & Hongfang Zhang. (2000). Synthesis and Crystalline Structure of Oligomer of Arylesterketone. *Chem Res, 11*, 5–7.

411. Guo Qingzhong, & Chen Tianlu. (2004). Synthesis of Macrosyclic Oligomers of Aryleneketones Containing Phthaloyl Links by Measn of Friedel-Crafts Acylation Reaction Chem. Lett, *33(4)*, 414–415.

412. Wang Hong Hua, Ding Jin, & Chen Tian Lu. (2004). Cyclic Oligomers of Phenolphthalein Polyarylene Ester Sulfone (ketone): Preparation Through Cyclo-Depolymerization of Corresponding Polymers. *Chin. Chem. Lett, 15(11)*, 1377–1379.

413. Kharaev, A. M., Basheva, R. Ch, Istepanova, O. L., Istepanov, M. I. & Kharaeva, R. A. (2006). Aromatic Oligoesterketones for Polycondensation RF Patent 2327680. International Paten Catalogue C 07 C 43/02.

414. Bedanokov A. Yu, Shaov A. Kh, Kharaev A. M. & Dorofeev V. T. (2000). *Plast. Massy, 4*, 42.

415. Bedanokov Azamat, U., Shaov Abubekir, Ch, Charaev Arsen, M. & Mashukov Nurali, I. (1997). *Sythesis and Some Properties of Oligo-and Polyesterketones Based on Bisphenylpropane*/in Proceedings of International Symposium New Approaches in Polymeric Syntheses and Macromolecular Formation. Sankt-Petersburg, 13–17. [in Russian].

416. Kharaev, A. M., Bazheva, R. Ch, Kazancheva, F. K., Kharaeva, R. A., Bahov, R. T., Sablirova, E. R. & Chaika, A. A. (2005). *Aromatic Polyesterketones and Polyesteresterketones as Perspective Thermostable Constructional Materials*/in Proceedings of the 2-nd Russian Research-Practical Conference. Nalchik, 68–72. [in Russian].

417. Kharaev, A. M., Bazheva, R. Ch, Kharaeva, R. A., Beslaneeva, Z. L., Pampuha, E. V. & Barokova, E. B. (2005). *Producing of Polyesterketones and Polyesteresterketones on the Basis of Bisphenols of Various Composition*/in Proceedings of the 2-nd Russian Research-Practical Conference. Nalchik, 44–47. [in Russian].

418. Bazheva, R. Ch, Kharaev, A. M., Olhovaia, G. G., Barokova, E. B. & Chaika, A. A. (2006). Polyester-Polyesterketone Block-Copolymers/in Abstracts of the International Conference on Organic Chemistry "Organic Chemistry from Butlerov and Belshtein Till Nowadays." Sankt-Petersburg, 716. [in Russian].

419. Aromatic Polymers. GB Patent 1563223.

420. Polysulfoneesterketones Germany Patent Application 3742445.

421. Germany Patent Application 3742264.

422. Aromatic Polymers. *Macromolecules*, 1984, *17(1)*, 10–14.

423. Khasbulatova, Z. S., Kharaev, A. M. & Mikitaev, A. K. (2009). Khim. *Prom. Segodnya, (10)*, 29–31.

424. Wen Hong-Li, Song Cai-Sheng, Tong Yong-Fen, Chen Lie, & Liu Xiao-Ling. (2005). Synthesis and Properties of Poly(aryl ester sulfone ester ketone ketone) (PESEKK) *J. Appl. Polym. Sci, 96(2)*, 489–493.

425. Li Wei, & Cai Ming-Zhong Ying. (2004). *Chin. J. Appl. Chem, 21(7)*, 669–672.

426. Sheng Shou-Ri, Luo Qiu-Yan, Yi-Huo, Luo Zhuo, Liu Xiao-Ling, & Song Cai-Sheng. (2008). Synthesis and Properties of Novel Organosoluble Aromatic Poly(ester ketone)s Containing Pendant Methyl Groups and Sulfone Linkages *J. Appl. Polym. Sci, 107(1),* 683–687.

427. Tong Yong-Fen, Song Cai-Sheng, Wen Hong-Li, Chen Lie, Liu Xiao-Ling. (2005). Synthesis and Properties of Copolymers Containing Methyl Replacers Polym. *Mater. Sci. Technol, 21(2),* 162–165.

428. Sheng Shou-Ri, Luo-Qiu-Yan, Huo Yi, Liu Xiao-ling, Pei Xue-liang, & Song Cai-sheng. (2006). Synthesis and Properties of Soluble Methyl-Replaced Polyarylesterketonestersulfonesterketones. *Polym. Mater. Sci. Technol, 22(3),* 85–87, 92.

429. Xie Guang-Liang, Liao Gui-Hong, Wu Fang-Juan, & Song Cai-Sheng. (2008). Synthesis and Adsorption Properties of Poly(arylestersulfonesterketone)Ketone with Lateral Carboxylic Groups. *Chin. J. Appl. Chem, 25(3),* 295–299.

430. Charaev, A. M., Khasbulatova, Z. S., Basheva, R. Ch, Kharaeva, R. A., Begieva, M. B., Istepanova, O. L. & Istepanov, M. I. (2007). *Izv. Vuzov. Sev.-Kav. Reg. Estestv. Nauki, 3,* 50–52.

431. Chen Lie, Song Cai-Sheng, Wen Hong-Li, Tong Yong-Fen, & Liu Xiao-Ling. (2004). Synthesis of Statistical Polyestersulfonesterketoneketones Containing Bis(o-methyl) Groups. *Chin. J. Appl. Chem, 21(12),* 1245–1248.

432. Arthanareeswaran, G., Mohan, D. & Raajenthiren, M. (2007). Preparation and Performance of Polysulfone-Sulfonated Poly(ester ester ketone) Blend Ultrafiltration Membranes. Part I. *Appl. Surface Sci, 253(21),* 8705–8712.

433. Xing Peixiang, Robertson Gilles, P., Guiver Michael, D., Mikhailenko Serguei, D. & Kaliaguine Serge. (2004). Sulfonated Poly(aryl ester ketone)s Containing Naphthalene Moieties for Proton Exchange Membranes. *J. Polym. Sci. A, 42(12),* 2866–2876.

434. Khasbulatova, Z. S. & Shustov, G. B. (2009). *Aromatic Oligomers for Synthesing Polyesters.* In Proceedings of the X International Conference of Chemistry and Physicochemistry of Oligomers. Volgograd, 99. [in Russian].

435. Khasbulatova, Z. S., Shustov, G. B. & Mikitaeva, A. K. (2010). *Vysokomol. Soed. B, 52(4),* 702–705.

CHAPTER 4

MICROHETEROGENEOUS TITANIUM ZIEGLER-NATTA CATALYSTS: 1,3-DIENE POLYMERIZATION UNDER ULTRASOUND IRRADIATIONS

VADIM P. ZAKHAROV, VADIM Z. MINGALEEV,
IRIVA D. ZAKIROVA, and ELENA M. ZAKHAROVA

CONTENTS

ABSTRACT

Polymerization of butadiene and isoprene under action of microheterogeneous titanium based catalyst with ultrasonic irradiation of the reaction mixture at the initial time is studied. It is shown that ultrasonic irradiation causes transformation multisite catalyst in a quasi-single site. At that activity of dominant site depends on the monomer nature.

4.1   INTRODUCTION

One of the important application area of ultrasound (US) is catalytic reactions with the participation of low molecular mass compounds and heterogeneous catalysts. The effect of ultrasound on catalytic reactions in the presence of platinum and rhodium catalysts of various dispersities was investigated in Ref. [1]. It was demonstrated that ultrasound can provide for the occurrence of chemical processes that cannot be performed even in the presence of catalysts. It is assumed that the main mechanism of its action on catalytic processes consists in the dispersion of catalyst particles; however, as was shown in Ref. [1], the adhesion of particles can occur during the action of the so-called Bjerknes forces, that is, forces that promote the attraction of particles (primarily small particles) to a deformed bubble followed by their sticking together. As a consequence, the diffusion of reagents to the surface of a particle becomes more pronounced and the rate of the process increases.

It seemed useful to examine the effect of ultra sound on the catalytic polymerization of dienes in the presence of Ziegler–Natta titanium catalysts, because the catalytic system is microheterogeneous and potentially susceptible to the effect of ultrasound. Moreover, although a long time has passed since the development of techniques for the synthesis of stereoregular polydienes, these catalysts or rare-earth compounds with close catalytic mechanisms are in current use for the majority of large tonnage manufacturing of synthetic rubber. The large body of research on the synthesis of stereoregular polydienes and the absence of substantial changes in the organization of their commercial production are due to the lack of understanding of some specific features of Ziegler catalysis: in particular, the origin of the multisite nature and the relationship of the reactivity distribution of active sites to the stereoregulating activity, to the microheterogeneity of a catalyst, to the kinetic parameters of the process, etc. Thus, US irradiation, which can simultaneously affect several parameters of the process, is a useful tool for the study of specific features of polydiene synthesis with Ziegler–Natta catalysts.

The aim of this study is to investigate the polymerization of butadiene and isoprene in the presence of the catalytic system $TiCl_4$-Al $(iso-C_4H_9)_3$ under US irradiation of the reaction mixture during its formation.

4.2　EXPERIMENTAL PART

The reaction mixture was exposed to US irradiation on a UZDN-2T apparatus equipped with a conical irradiator operating at a frequency of 22 kHz; the maximum power was 400 W at a current strength of 25 mA. A 500 cm^3 reaction flask was hermetically connected to an ultrasound irradiator to prevent contact of the reaction mixture with the atmosphere.

The titanium catalytic complex was prepared through pouring of toluene solutions of TiCl4 and Al(iso-C4H9)3 into a separate reaction vessel. Then, the catalytic complex was aged at 0°C under stirring for 30 min to achieve the maximum activity. The optimum ratio of the catalytic system components, Al/Ti, was dependent on the nature of the monomer; these ratios were 1.4 and 1.1 for butadiene and isoprene, respectively.

Polymerization was performed in toluene free of trace moisture and admixtures that deactivate the catalyst. The temperature of polymerization was 25(±1)°C. The catalyst concentration was 5 mmol/L, and the monomer concentration was 1.5 mol/L. These values correspond to the maximum activities and stereo specificities of the catalyst for these monomers. Polymerization was conducted with the use of two methods.

The preliminarily prepared titanium catalyst was added to the monomer solution in a flow of argon. This time was taken as the onset of polymerization. The process was performed under constant stirring with a magnetic stirrer. The synthesis was conducted in a manner similar to that described above, but at the time of catalyst addition, the reaction mixture was subjected to ultrasonic stirring during the first minute of polymerization. Published data show that US irradiation can initiate the polymerization of some monomers in the absence of initiating agents. To estimate the contribution of ultrasound initiated polymerization, US irradiation of toluene solutions of the monomers (butadiene and isoprene) was performed for 1 min in the absence of the titanium catalytic complex. Then, polymerization was performed under constant stirring with a magnetic stirrer. The polymer was sampled directly from the reaction mixture in a flow of argon. To stop polymerization at the predetermined time of synthesis, a calculated amount of methanol, which facilitates decomposition of the catalytic complex and precipitation of the polymer from toluene solutions, was added to the reaction mixture. The residual catalyst was removed from the polymer via repeated washing of the sample with methanol (pH ~ 5–6). At the final stage, the polymer was reprecipitated and the samples were dried in vacuum to a constant weight. The yields of the polymers were determined gravimetrically.

The molecular-mass characteristics of the polymers were determined via GPC on a Waters GPC-2000 instrument. Calibration was made relative to narrowly dispersed polystyrene standards.

The microstructure of polybutadiene was studied via IR spectroscopy on a Shimadzu IR Prestige spectrometer. Analysis was performed with the use of polymer films applied on KBr glasses. The films were cast from toluene solutions. The microstructure of polyisoprene was studied via ^{1}H NMR spectroscopy on a Bruker AM-300 spectrometer. Deutero chloroform was used as a solvent.

4.3   RESULTS AND DISCUSSION

When the toluene solutions of isoprene and butadiene were US irradiated without any catalyst, ultrasound did not initiate polymerization processes under the given experimental conditions. After irradiation of monomer solutions and further stirring with a magnetic stirrer, even trace amounts of the polymers were not detected, as demonstrated by the absence of characteristic turbidity of solution upon the addition of methanol.

In the absence of ultrasonic irradiation of the reaction mixture (Method 1), polybutadiene with a wide polymodal molecular-mass distribution is formed at initial conversions; in contrast, polyisoprene is characterized by a narrower unimodal molecular-mass distribution (Fig. 4.1). During ultrasound treatment, the molecular-mass distribution of polybutadiene formed at initial conversions becomes significantly narrower owing to reduction in the shares of both low and high molecular mass fractions (Fig. 4.1). Under similar conditions, the width of the molecular-mass distribution of polyisoprene decreases slightly (Fig. 4.1) in this case Mw of the polymer tend to increase (Fig. 4.2). Thus, regardless of the nature of a monomer, US irradiation facilitates the synthesis of polybutadiene and polyisoprene with narrower molecular-mass distributions. Since the statistical width of the molecular-mass distributions for polybutadiene and polyisoprene synthesized via methods 1 and 2 is greater than the polydispersity index for the most probable Flory distribution, it is reasonable to suggest that the titanium catalytic system is kinetically heterogeneous.

In terms of the multisite nature of Ziegler–Natta catalytic systems [2, 3], each functioning type of active site generates polymer fractions with a certain molecular mass and stereoregular composition; i.e., polymerization sites are kinetically and stereospecifically heterogeneous.

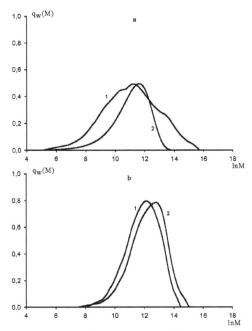

FIGURE 4.1 Molecular-mass distributions of (a) polybutadiene and (b) polyisoprene: (*1*) Method 1 and (*2*) Method 2. Conversion is 1–3%.

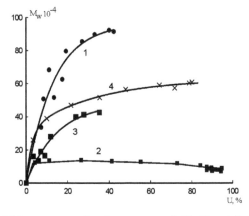

FIGURE 4.2 Weight-average molecular masses of (*1, 2*) polybutadiene and (*3, 4*) polyisoprene vs. conversion: (*1, 3*) Method 1 and (*2, 4*) Method 2.

On the basis of experimental molecular-mass distributions of the poly-mers, the inverse task was solved via Tikhonov's regularization method [3]; as a result, the functions of distribution of active sites over the probability of chain termination were obtained (Fig. 4.3, 4.4). The titanium catalytic system is characterized by a polymodal distribution function, where each maximum corresponds to a certain type of polymerization sites. Under our conditions of polybutadiene synthesis, there are four types of active sites that form macro-molecules with the following most probable molecular masses: $\ln M = 9.2$–10.4 (I), 11.2–11.4 (II), 12.9–13.2 (III), and 14.1–14.7 (IV) (Fig. 4.3). The position of maxima on the curves is conversion independent. This implies that each type of site forms macromolecules of a certain length throughout the process.

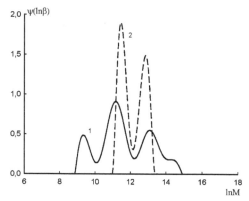

FIGURE 4.3   Kinetic-heterogeneity distributions of active sites for polymerization of butadiene: (*1*) Method 1 and (*2*) Method 2. Conversion is 1–3%.

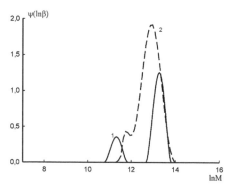

FIGURE 4.4   Kinetic heterogeneity distributions of active sites for polymerization of isoprene: (*1*) Method 1 and (*2*) Method 2. Conversion is 1–3%

When ultrasound treatment was used at the onset of polymerization of butadiene, the titanium catalytic system featured only type II and type III polymerization sites, as evidenced by the occurrence of only two maxima on the curve of the distribution of active sites over kinetic heterogeneity (Fig. 4.3). A decrease in the types of sites during polymerization via Method 2 is responsible for narrowing of the molecular-mass distribution of polybutadiene.

During the synthesis of polybutadiene via Method 1, the activities of type I and type II sites, which generate polybutadiene fractions with relatively low molecular masses, decrease during polymerization (Fig. 4.5). In contrast, the activities of type-III and type IV sites (Fig. 4.5) that form a high molecular mass polydiene increase with monomer conversion.

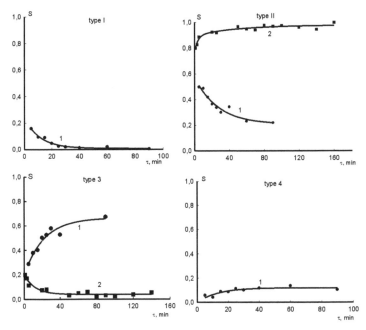

FIGURE 4.5 Variation in the kinetic active sites of butadiene polymerization: (*1*) Method 1 and (*2*) Method 2.

UV irradiation brings about an abnormal change in the activity of functioning types of polymerization sites. The activity of type II polymerization sites tends toward unity (Fig. 4.5); that is, the multisite titanium catalytic system is transformed into a quasi single site system during ultrasonic treatment. Thus already at 3 min of polymerization, the fraction of the monomer polymerized on propagating type II sites achieves 90%. When the time of polymerization is increased to 60 min, this fraction of the monomer becomes as high as 98%.

Solution of the inverse problem of formation of the molecular-mass distribution for the polymerization of isoprene showed that, when polymerization was performed via Method 1, the titanium catalytic system was characterized by the presence of type II and type III active sites (Fig. 4.4).

In this case, active sites forming high-molecular mass polyisoprene show higher activities than those in the polymerization of butadiene. The kinetic activity of type II sites decreases during the polymerization of isoprene under US irradiation (Fig. 4.6). In this case, the activity of type III sites increases significantly (Fig. 4.7). Thus, at a time of polymerization of 50 min, the fraction of isoprene polymerized on type III sites is as high as 95%.

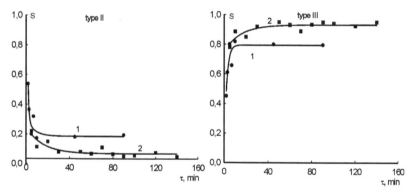

FIGURE 4.6   Variation in the kinetic active sties of isoprene polymerization: (*1*) Method 1 and (*2*) Method 2.

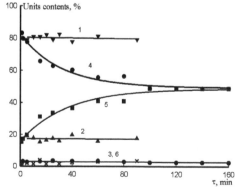

FIGURE 4.7   Contents of (*1, 4*) cis-1,4-, (*2, 5*) trans1,4-, and (*3, 6*) 1,2-units in polybutadiene: (*1–3*) Method 1 and (*4–6*) Method 2.

After US irradiation of the reaction mixture, the fraction of cis-1,4-units in polybutadiene decreases, while the fraction of trans1,4-units increases with an increase in conversion (Fig. 4.7). Polybutadiene formed at high monomer conversions contains equal amounts of cis and trans units (48.8% trans1,4-, 48.8% cis-1,4-, and 2.4% 1,2-units). US treatment has no effect on the content of 1,2-units in polybutadiene. At the same time, ultrasound mixing does not alter the cis stereo specificity of the titanium catalytic system in the polymerization of isoprene.

An examination of dispersity of the micro heterogeneous titanium catalyst in the absence of ultrasound showed that, for catalyst particles, the most probable weight-average radius is 4.2 μm (Fig. 4.8). In this case, the share of fractions of relatively large particles (6–12 μm) decreases. Thus, the ultrasound irradiation of the catalytic system leads to narrowing of the size distribution of particles without any noticeable change in the most probable radius. This effect is most probably associated with the fact that, during US mixing, the size distribution of particles is the result of tow processes: on the one hand, the dispersion of large catalyst particles during the action of ultrasound vibrations of the solid medium and, on the other hand, the adhesion of small catalyst particles during the action of Bjerknes forces.

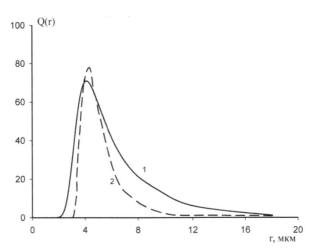

FIGURE 4.8   Radius distributions of titanium catalyst particles: (*1*) Method 1 and (*2*) Method 2.

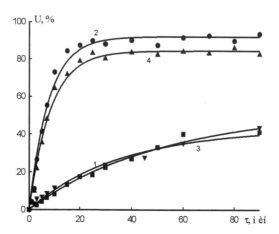

FIGURE 4.9 Yields of (*1, 2*) polybutadiene and (*3, 4*) polyisoprene vs. time of polymerization: (*1, 3*) Method 1 and (*2, 4*) Method 2.

The kinetic curves for the polymerization of butadiene and isoprene with the titanium catalytic system in the absence of US irradiation (Method 1) are almost coincident (Fig. 4.9). US irradiation (Method 2) brings about an increase in the initial rates of butadiene and isoprene polymerization and accelerates accumulation of the polymer in the system. In this case, the initial rates of polymerization of butadiene and isoprene increase owing to an increase in the rate constant of chain propagation without any marked changes in the concentration of active sites. This result correlates with the estimation of dispersity of the catalytic system, specifically, with the absence of changes in the most probable size of catalyst particles. In the case of polymerization of butadiene via Method 2, the numerical values of rate constants of chain termination increase.

4.4   CONCLUSION

The ultrasonic irradiation of the reaction mixture at the initial time of the polymerization of butadiene and isoprene under action Ti-Al catalyst result in an increase of process rate. Polymerization occurs at the single active site reactivity of which depends on the nature of diene. With ultrasonic treatment the reactivity of the active sites of polymerization of isoprene greater than for the butadiene polymerization. Reduction of polymerization reactivity site butadiene reduces cis-specificity titanium catalyst. In all cases ultrasound irradiation produces a polydienes with a narrow molecular weight distribution.

ACKNOWLEDGMENTS

This study was financially supported by the Council of the President of the Russian Federation for Young Scientists and Leading Scientific Schools Supporting Grants (project no. MD-4973.2014.8).

KEYWORDS

- **Diene polymerization**
- **Heterogeneous catalysis**
- **Titanium Ziegler-Natta catalyst**
- **Ultrasound**

REFERENCES

1. Margulis, M. A. (1986). *Sonochemical Reactions and Sonoluminescence*, Moscow: Khimiya, 286 p.
2. Kissin, Yu. V. (2012). *Journal of Catalysis 292*, 188–200.
3. Monakov, Y. B., Sigaeva, N. N. & Urazbaev, V. N. (2005). *Active Sites of Polymerization. Multiplicity: Stereospecific and Kinetic Heterogeneity*, Leiden: Brill Academic, 397 p.

CHAPTER 5

ON ELECTRIC CONDUCTIVITY OF POLYMER COMPOSITES: A CASE STUDY

J. N. ANELI, T. M. NATRIASHVILI, and G. E. ZAIKOV

CONTENTS

ABSTRACT

On the basis of mixes of phenolformaldehide and epoxy resins at presence of some silicon organic compounds and fiber glasses annealed in vacuum and hydrogen media the new conductive monolithic materials have been created. There were investigated the conductive, magnetic and some other properties of these materials. It is established experimentally that the obtained products are characterized by semiconducting properties, the level of conductivity of which are regulated by selection of technological conditions. The density and mobility of charge carriers increase at increasing of annealing temperature up to definite levels. The temperature dependence of the electrical conductivity and charge mobility describe by Mott formulas. It is established that at annealing free radicals and other paramagnetic centers are formed. It is proposed that charge transport between conducting clusters provides by mechanism of charge jumping with alternative longevity of the jump.

5.1   INTRODUCTION

Interest to the processes proceeding at increased temperatures (up to 600K) in polymer materials is stimulated by a possibility to obtain systems with double conjugated bonds, which exhibits the properties of semiconductors. Intensity of formation of the polyconjugation systems increases with temperature, if pyrolysis proceeds in vacuum or under inert atmosphere[1, 2].

Intramolecular transformations with further change of supermolecular system were studied well on the example of polyacrylonitrile [3]. Thermal transformation of polyacrylonitrile leads to formation of a polymer, consisted of condensed pyridine cycles with conjugation by C=C bonds, as well as by C=N ones. Concentration of paramagnetic particles increases with temperature of pyrolysis. It is known that the ESR signal is one of the signs of polyconjugation appearance in polymer systems [4]. Deep physical and chemical transformations in polymers proceed at combination of temperature varying with introduction of various donor-acceptor inorganic or organic additives into the reactor.

It is known that one of the ways of obtaining of organic semiconductors is the pyrolysis of low or high molecular substances. However, as a rule, this method allows a formation of powder like materials and formation of monolithic ones is connected with additive technological procedures, after which often the material deteriorated.

The main aim of the presented work is the obtaining of pyrolyzed monolithic materials with wide range of the electric conductivity.

5.2 EXPERIMENTAL PART

Epoxy resin (ER), novolac phenoloformaldehyde resin (PFR), polymethyl-silsesquioxane (PMS) and fiber glass (FG) were chosen as the initial substances. Pyrolysis of mixtures of the components mentioned, pressed in press-forms, was conducted at various temperature ranged within 500–1500 K in 10 Pa vacuum. Products obtained in this manner possess good mechanical and electroconducting properties, and are monolythic materials. Pyrolized samples or pyrolyzates were tested by polarization microscopy technique in order to determine their microstructure. The paramagnetic properties of pyrolizates were investigated by using of ESR spectrometer of Brukker type. The type and mobility of charge carriers investigated were measured by the Hall effect technique.

The main aim of this chapter is to obtaining pyrolyzed monolithic materials with wide range of the electric conductivity.

5.3 RESULTS AND DISCUSSION

Inclusion of fiber glass into compositions was induced by the following idea. It is known [5] that at high temperatures organosiloxanes react with side hydroxyl groups, disposed on the fiber glass surface. In this reaction they form covalent bonds with those side groups.

It is known that after high-temperature treatment silsesquioxanes obtain a structure, close to inorganic glass with spheres of regulation due to formation of three-dimension siloxane cubic structures and selective sorption of one of the composite elements is possible on the filler surface in the hardening composite [5].

Figures 5.1–5.3 reflect changes of some mechanical, electric and paramagnetic properties of polymer composites depending on pyrolysis temperature. These dependences are the result of proceeding of deep physicochemical transformations in materials. Combined analysis of the change of microstructure and density of materials (Fig. 5.1) with the increase of pyrolysis temperature induces a conclusion that excretion of some volatile fractions of organic part of the material, carbonization of organic residue and caking of glassy fibers cause the increase of pyrolyzate density, based on the composite with polymethylsesquioxane. The limit of pyrolyzate density is reached at temperatures near1273 K (Fig. 5.1, curve 1), followed by a decrease of the material density due to intensification of thermal degradation processes with pyrolysis temperature increase above 1273 K.

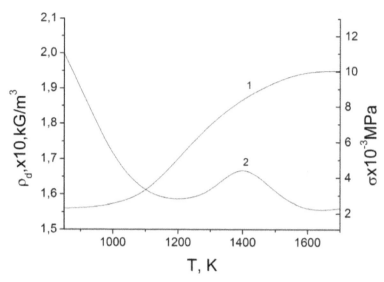

FIGURE 5.1   Dependences of ρ_{den} (1) and strengthening at elongation s (2) on pyrolysis temperature for the composite ED-20 + PFR + KO-812 + FG.

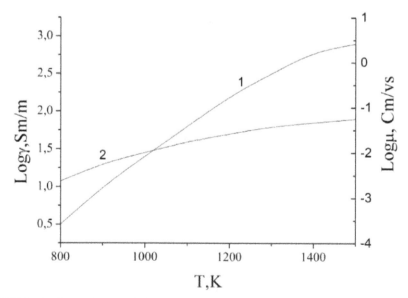

FIGURE 5.2   Dependences of electric conductivity g (1), mobility of charge carriers m (2) on pyrolysis temperature for the composite ER + PFR + PMS + FG.

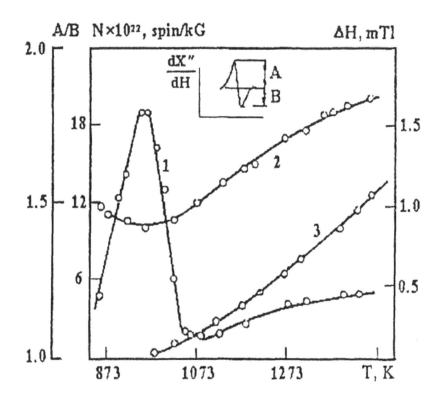

FIGURE 5.3 Temperature dependence of paramagnetic centers N (1), ESR line width (2) and ESR line asymmetry parameter A/B (3) on the pyrolysis temperature for the ED-20 + PFS + KO-812 + FG composites.

The material strengthening at elongation extremely depends on the pyrolysis temperature, possessing an intermediate maximum near 1273 K (Fig. 5.1, curve 2). Burning out of organic part of the composite leads to weakening of adhesion forces in the interphase and, consequently, to decrease of the material strengthening with pyrolysis temperature increase up to a definite value. At further increase of pyrolysis temperature on the curve of this dependence the small maximum appears due formation of covalent chemical bonds between glass and organic conjugated double bounds skeleton. At more high temperatures of pyrolysis the degradation of these bonds has place

The conductivity (g) and charge carrier mobility of the pyrolyzates grows monotonously initially with increasing of the pyrolysis temperature and then saturate. This dependence points out a constant accumulation of polyconjugation systems due to complex thermochemical reactions. Chemical bonds which link organic and inorganic parts of the composite reliably increase stability of polyconjugated structures, responsible for electrically conducting properties of materials. The electrically conducting system of the materials can be considered as a heterogeneous composite material, consisted of highly conducting spheres of polyconjugation and barrier inter layers between them.

The most apparently true model of electric conductivity in materials with the system of double conjugated bonds seems to be the change transfer in the ranges of polyconjugation possessing metal conductivity and jump conductivity between polyconjugation spheres [6].

An important information on the nature of conductivity of pyrolyzed polymer materials is given by investigation of the g dependence on temperature. Comparison of the experimental data on dependence of g-T with known for organic semiconductors [2]:

$$g = \gamma_0 \exp\left(-dE/kT\right) \tag{1}$$

and one proposed by N. Mott shows that the dependence obtained by us experimentally satisfies to Mott low [7]:

$$\gamma = \gamma_0 \exp\left[-\left(\frac{T_0}{T}\right)^{1/4}\right], \tag{2}$$

where T_0 and γ_0 are constants depending on some quantum mechanical values.

The growth of carriers mobility m well described with analogical expression:

$$\mu = \mu_0 \exp\left[-\left(\frac{T_0}{T}\right)^{1/4}\right] \tag{3}$$

The dependence of paramagnetic centers concentration in pyrolyzed polymer composites on pyrolysis temperature has an extreme character (Fig. 5.3). Curve of the present dependence possesses maximum, which is corresponded to the 900–1000 K range. Change of the ESR absorption line intensity is accompanied by a definite change of its width. In this case, the form and width of the ESR line changes (at constancy of the g-factor)-lines are broadened, and asymmetry of singlet occurs. Maximum on the concentration dependence for paramagnetic centers on pyrolysis temperature is correspondent to the temperature range, in which volatile products of pyrolysis are released and polyconjugation systems occur. Decrease of concentration of the centers above 973 K proceeds due to coupling of a definite amount of unpaired electrons. According to this coupling new chemical structures occur (for example, polyconjugation responsible for electric conductivity increase).

At more high temperatures of pyrolysis deepening of thermo-chemical reactions in composites leads to formation of the paramagnetic centers localized on the oxygen atom. On the other hand, it is probable of the increase of free charges-current carriers contribution into ESR signal, the line of which is characterized by asymmetry (so called Dayson form [8]).

5.4   CONCLUSIONS

High-temperature treatment (pyrolysis) of polymer composites in the inert atmosphere or in the hydrogen medium stimulates processes of formation of the polyconjugation systems). Charge transfer between polyconjugation systems is ruled by the jump conductivity mechanism with variable jump length. In this case, its temperature dependence is described by the Mott formulas. Presence of a glassy fiber and polymethylsilasesquioxane in composites promote formation of covalent bonds between organic and inorganic parts of the composite at pyrolysis. This leads to improving of mechanical properties of materials together with the electric ones.

KEYWORDS

- **Charge transport**
- **Conductivity**
- **Electric**
- **Paramagnetic centers**
- **Phenolformaldehide and epoxy resins**
- **Pyrolysis**
- **Silicon organic compound**

REFERENCES

1. Brütting, W. (2005). Physics of Organic Semiconductors. Wiley-VCH.
2. Aviles, M. A., Gines, J. M., Del Rio, J. C., Pascual, J., Perez, J. L. & Sanchez-Soto, P. J. (2002). *J. Thermal Analysis and Calorimetry, 67,* 177–188.

3. Fialkov, A. S. (1979). Carbon-Graphite Materials, Moscow, Energia, 158.
4. Milinchuk, V. K., Klinshpont, E. R. & Pshezhetskii, S. Y. (1980). Macroradicales, Moscow, Khimia, 264.
5. Aneli, J. N., Khananashvili, L. M. & Zaikov, G. E. (1998). "Structuring and Conductivity of Polymer Composites," *Nova Sci.* Publ. N.-Y., 326 P.
6. Kajiwara, T., Inokuchi, H., Minomura, S. (1974). *Jap. Plast. Age, 12(1),*17–24.
7. Mott, N. F. & Davis, E. (1979). Electron Processes in Noncrystalline Materials, 2nd Ed, Oxford, Clarendon Press.
8. Dyson, F. J. (1955). *Phys. Rev, 98,*349–358.

CHAPTER 6

A STUDY ON RUBBER VULCANIZATES CONTAINING MODIFIED FILLERS

DARIUSZ M. BIELIŃSKI, MARIUSZ SICIŃSKI, JACEK GRAMS, and MICHAŁ WIATROWSKI

CONTENTS

6.1   INTRODUCTION

Powders are commonly used as fillers for rubber mixes. The most popular are carbon black, silica, kaolin, or more modern like graphene, fullerenes and carbon nanotubes. The nature of their surface is the main attribute of fillers, as surface energy and specific area determine the compatibility of filler with rubber matrix and the affinity to other c ingredients. One of the major problems is the tendency of fillers to agglomeration–formation of bigger secondary structures, associated with lower level of filler dispersion, what is reflected by the decrease of mechanical properties of rubber vulcanizates [1]. Surface modification of powder can improve interaction between rubber matrix and filler. Application of low-temperature plasma treatment for this purpose has been drown increasing attention recently [2, 3].

Silica is one of the most popular mineral filler used in rubber technology. Three types of silica can be distinguished and namely: precipitated, fumed and surface-modified silica. As an amorphous material with randomly placed functional silanol groups (Fig. 6.1), it readily generates hydrogen bonds with surrounding molecules [4].

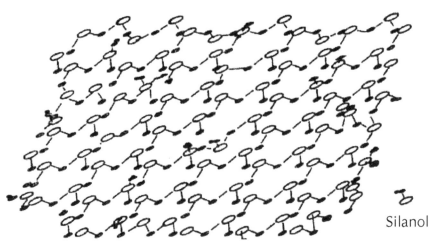

Silanol

FIGURE 6.1   Surface chemistry of silica [4].

Polar character and big specific surface area enable various modifications of silica surface. Modifying by coupling compounds is the most popular one [4]. In subject literature [5, 6] and patent declarations [7] many references on the modification processes, their kinetics, current opportunities and proposals of further development, can be found. All chemical methods have a significant disadvantage: emission of large amounts of chemical waste, usually in the form of harmful solvents.

Taking into account the necessity of their utilization, application of "clean" plasma modification has to be considered as a cost effective possibility for significant reduction of environmental hazard.

Low-temperature plasma can be generated with a discharge between electrodes in a vacuum chamber. The process used to be carried out in the presence of gas: that is, Ar_2, O_2, N_2, methane or acetylene. Depending on the medium applied, surface of modified material can be purified, chemically activated, or grafted with various functional groups.

This paper presents the results of low-temperature, oxygen plasma activation of silica, kaolin and wollastonite. Fillers were modified in a tumbler reactor, enabling rotation of powders in order to modify their entire volume effectively. Based on our previous work [8], the process was carried out with 100W discharge power. The time of modification varied from 8 to 64 min. Additionally, for the most favorable (in terms of changes to surface free energy) time of modification for kaolin, the process was repeated and ended with a flushing of the reactor chamber with hydrogen, in order to reduce of carboxyl groups content, generated on filler surface. Rubber mixes, filled with the modified powders, based on SBR or NBR were prepared and vulcanized. Mechanical properties of the vulcanizates were determined and explained from the point of view rubber-filler interactions and filler, estimated from micro morphology of the materials.

6.2 EXPERIMENTAL PART

6.2.1 MATERIALS

6.2.1.1 RUBBER VULCANIZATES

Three fillers were the objects of study: micro silica Arsil (Z. Ch. Rudniki S.A., Poland), kaolin KOM (Surmin-Kaolin S.A., Poland) and wollastonite Casiflux (Sibelco Specialty Minerals Europe, The Netherlands). Rubber mixes, prepared with their application, were based on: styrene-butadiene rubber (SBR) KER 1500 (Synthos S.A., Poland) and acrylonitrile-butadiene rubber (NBR) NT 1845 (Lanxess, Germany).

Rubber mixes were prepared with a Brabender Plastic order laboratory micro mixer (Germany), operated with 45 rpm, during 30 min. Their composition is presented in Table 6.1. The only one variable was the type of modified mineral filler (see Section 6.2.1).

TABLE 6.1   Composition of the Rubber Mixes Studied

Components	Content [phr]	
SBR KER1500	100	0
NBR NT1845	0	100
ZnO	5	5
Stearine	1	1
CBS	2	2
Sulfur	2	2
Arsil Silica	20	20
Modified filler	20	20

Samples were vulcanized in 160 °C, time of vulcanization: 6 min. (for NBR vulcanizates) and 15 min. (for SBR vulcanizates).

Symbols of the prepared vulcanizate samples:

- NBR-X-composite based on NBR rubber, X-modified filler;
- SBR-X-composite based on SBR rubber, X-modified filler.

6.2.1.2   PLASMOCHEMICAL MODIFICATION OF FILLERS

Fillers studied were modified with a Diener tumbler plasma reactor (Germany). The reactor operated with the frequency of 40 kHz and the maximum discharge power of 100 W. Scheme of the reactor is shown in Fig. 6.2.

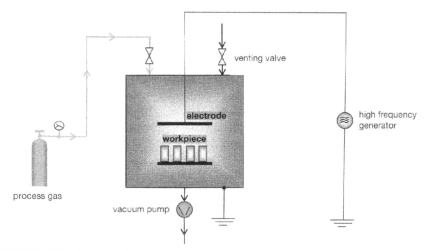

FIGURE 6.2   Scheme of the plasma reactor.

Mineral fillers were subjected to the oxygen plasma treatment during various time. Efficiency of process gas flow was 20 cm³/min, and the pressure in the reactor chamber was maintain at 30 Pa. Symbols of the modified fillers are as follows:

- A-REF–silica Arsil, virgin reference filler;
- A-XX–silica Arsil, modified during time of XX min (XX = 16; 32; 48; 64);
- K-REF–kaolin, virgin reference filler;
- K-XX–kaolin, modified during time of XX min (XX = 8; 16; 32);
- K-16H–kaolin, modified during 16 min, the process terminated with hydrogen;
- W-REF–wollastonite, virgin reference filler;
- W-XX–wollastonite, modified during time of XX min (XX = 16; 32; 48; 64).

6.2.2 TECHNIQUES

6.2.2.1 SURFACE FREE ENERGY OF FILLERS

Effectiveness of plasmochemical modification of the fillers is represented by changes to their surface free energy (SFE) and its components–polar and dispersion one. SFE was examined with a K100 MKII tensiometer (KRÜSS GmbH, Germany). Contact angle was determined using polar (water, methanol, ethanol) and nonpolar (n-hexane, n-heptane) liquids. SFE and its components were calculated by the method proposed by Owens-Wendt-Rabel-Kaeble [9].

6.2.2.2 MICROMORPHOLOGY OF RUBBER

Micro morphology of rubber vulcanizates was studied with an AURIGA (Zeiss, Germany) scanning electron microscope (SEM). Secondary electron signal (SE) was used for surface imaging. Accelerating voltage of the electron beam was set to 10 keV. Samples were fractured by breaking after dipping in liquid nitrogen.

6.2.2.3 MECHANICAL PROPERTIES OF RUBBER VULCANIZATES

Mechanical properties of the vulcanizates studied were determined with a Zwick 1435 universal mechanical testing machine (Germany). Tests were carried out on "dumbbell" shape, 1.5 mm thick and 4 mm width specimens, according to PN-ISO 37:1998 standard. The following properties of the materials were determined: elongation at break (Eb), stress at elongation of 100% (SE100), 200% (SE200), 300% (SE300) and tensile strength (TS).

6.3   RESULTS AND DISCUSSION

Our previous studies revealed, that low-temperature plasma causes changes to surface free energy and its component of carbon nanotubes [10]. Plasma modification is a good method of CNT purification as an amorphous carbon is eliminated from their surface during process [11]. Purifying changes properties of CNT, and affects its dispersion in rubber matrix. It encouraged us to try plasma modification to silica, kaolin and wollastonite. The objective of the study was to characterize changes to filler surface and its susceptibility to oxygen activation, being expected to be an intermediate step in surface functionalization with various chemical groups/compounds.

6.3.1   SURFACE FREE ENERGY (SFE)

Reference silica powder represents relatively low value of surface energy and its polar component (Fig. 6.3a)–probably because of physically adsorbed water present on filler surface. After modification-regardless time of the process–SFE remains constant, however the dispersive component decreases in favor of the polar component increasing. Generally, silica remains resistant to plasma modification under the experimental conditions. However, it seems likely to change under higher discharge power.

Wollastonite behaves in a different way (Fig. 6.3b). After 48 min of plasma treatment value of its SFE reaches a maximum. Its polar component becomes almost doubled–probably because grafting of oxygen groups on filler surface. After 48 min of the treatment polar component of SFE is decreasing, probably because the process balance moves towards the surface cleaning.

Plasma treatment of kaolin (Fig. 6.3c) during 16 min results in an increase of SFE value and its polar component. After this time further changes are not observed. Hydrogen termination of the process (lasting 2 min) results in almost doubled the polar component of SFE, probably being the effect of surface present carbonyl groups reduction to the more stable carboxyl ones.

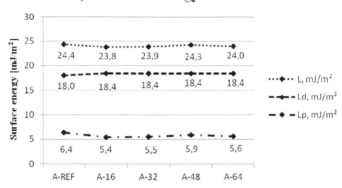

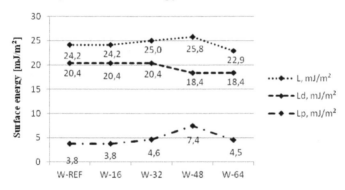

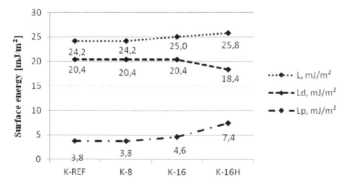

FIGURE 6.3   Results of the analysis of total surface free energy of fillers: (a) Silica, (b) Wollastonite, (c) Kaolin (L–total surface free energy, L_d–dispersive part, L_p–polar part).

6.3.2 MORPHOLOGY OF RUBBER VULCANIZATES

In order to determine the influence of rubber matrix polarity on filler dispersion and rubber-filler interaction, two kinds of rubber: NBR and SBR, were chosen. SEM pictures of the rubber vulcanizates, filled with reference and 48 min plasma treated wollastonite, are presented in Fig. 6.4. Morphology of SBR/wollastonite samples does not reveal any changes, explaining strengthening of the material (see the next section).

A   B

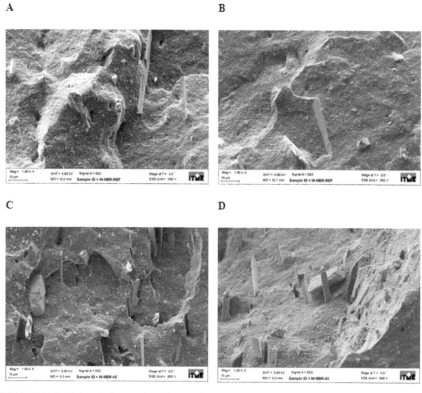

C   D

FIGURE 6.4   Morphology of rubber vulcanizates studied: A, B–NBR-W-REF; C, D–NBR-W-48; magnification 5000×.

Pictures of NBR-W-REF samples (Fig. 6.4a, b) present broken needles of wollastonite in the area of fracture, whereas in the case of NBR-W-48 sample (Fig. 6.4c, d) needles of wollastonite are nonbroken but "pulled out" from rubber matrix. This change to morphology, reflected by lower rubber-filler interactions, responsible for worse mechanical properties of rubber vulcanizates (see Section 6.3.3), is undoubtedly the result of an increase of SFE polar component of filler after plasma treatment. The SEM pictures of the vulcanizates, no matter, containing virgin or modifies wollastonite particles, do not reveal any filler agglomeration.

Morphology of SBR-K-REF and SBR-K-16H samples are presented in Fig. 6.5. Agglomerates of kaolin can be seen in vulcanizate containing reference filler (Fig. 6.6a, b). Modified kaolin does not exhibit tendency to agglomeration (Fig. 6.5c, d). Better filler dispersion suggests on higher mechanical properties of the SBR vulcanizates filled with plasma treated kaolin.

A   B

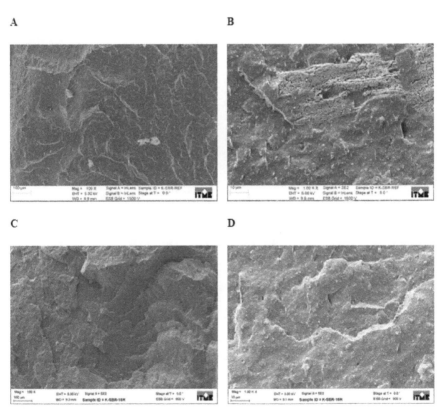

C   D

FIGURE 6.5 Morphology of rubber vulcanizates studied: A, B–SBR-K-REF; C, D–SBR-K-16H; magnification 100× (A, C) and 1000× (B, D).

6.3.3   MECHANICAL PROPERTIES OF RUBBER VULCANIZATES

Mechanical properties of the vulcanizates studied, containing virgin and plasma modified fillers are presented in Fig. 6.6a–f.

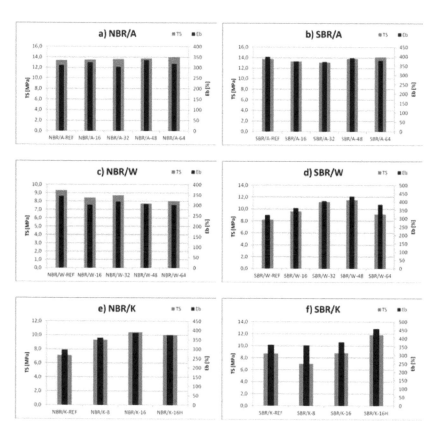

FIGURE 6.6   Mechanical properties of elastomer composites based on NBR and SBR filled with: (a)–(b) silica, (c)–(d) wollastonite, (e)–(f) kaolin; TS–tensile strength, Eb–elongation at break.

Plasma modification does not cause any changes to surface free energy of silica. This is clearly reflected by the mechanical properties of silica filled rubber vulcanizates (Fig. 6.6a, b). Changes to the values of material stress at elongation 100, 200 and 300% (SE100, SE200 and SE300), its tensile strength (TS) and elongation at break (Eb), being the result of plasma treatment of the filler, are negligible.

Mechanical properties of wollastonite filled NBR vulcanizates decrease due to plasma modification of the filler (Fig. 6.6c), whereas in case of vulcanizates based on SBR an increase of TS and Eb is observed (Fig. 6.6d)–especially for the most effective 48 min treatment. SEM pictures of the vulcanizates confirm on adequate changes to their morphology.

For the rubber vulcanizates filled with kaolin (Fig. 6.6e, f), despite the biggest changes to surface free energy and its components (observed for 16 min plasma treatment followed by hydrogen termination), determined changes to mechanical properties are different in comparison to the wollastonite filled vulcanizates. The biggest increase of TS and Eb is observed for SBR/K-16H sample-about 30% as compared to the reference sample (containing virgin filler). Reinforcement of rubber seems to be dependent on overlapping effects originated from rubber-filler interactions and dispersion of filler in rubber matrix.

6.4   CONCLUSIONS

1. Oxygen plasma treatment can activate surface of mineral fillers, by grafting oxygen groups on the surface.
2. The efficiency of the treatment depends on the filler, Changes to surface free energy and its components are observed for kaolin and wollastonite, whereas practically no energetic effect is present in the case of silica.
3. Any changes to filler particles SFE and its components effect on mechanical properties of rubber vulcanizates filled with the modified filler. Improvement of mechanical properties of the materials originates increased rubber-filler interaction and better dispersion of filler particles in rubber matrix.

ACKNOWLEDGMENT

The project was funded by the National Science Centre Poland (NCN) conferred on the basis of the decision number DEC-2012/05/B/ST8/02922.

KEYWORDS

- Low-temperature plasma
- Mineral fillers
- Rubber vulcanizates
- Surface modification

REFERENCES

1. Wolff, S. & Wang, J. (1992). Filler Elastomer Interactions. Part IV. The Effect of the Surface Energies of Fillers on Elastomer Reinforcement. *Rubber Chemistry and Technology 65*, 329–342.
2. Dierkes, W. K., Guo, R., Mathew, T., Tiwari, M., Datta, R. N., Talma, A. G., Noordemeer, J. W. M. & van Ooij, W. J. (2011). A Key to Enhancement of Compatibility and Dispersion in Elastomer Blends. *Kautschuk Gummi Kunststoffe 64*, 28–35.
3. Chityala, A. & van Ooij, W. J. (2000). Plasma Deposition of Polymer Films on Pmma Powders Using Vacuum Fluidization Techniques. *Surface Engineering 16*, 299–302.
4. Wang, M.-J. (1998). Effect of Polymer-Filler and Filler-Filler Interactions on Dynamic Properties of Filled Vulcanizates. *Rubber Chemistry and Technology 71*, 520–589.
5. Hair, M. L. & Hertl, W. (1971). Reaction of Chlorosilanes with Silica. *Journal of Physical Chemistry 14*, 2181–2185.
6. Blume, A. (2011). Kinetics of the Silica-Silane Reaction. *Kautschuk und Gummi Kunststoffe*, 4, 38–43,
7. Revis, A. (2003). Chlorosilane Blends for Treating Silica. *US Patent* 6613139B1.
8. Bieliński, D., Parys, G. & Szymanowski, H. (2012). Plazmochemiczna Modyfikacja Powierzchni Sadzy Jako Napełniacza Mieszanek Gumowych. *Przemysł Chemiczny 91*, 1508–1512.
9. Owens, D. K. & Wendt, R. C. (1969). Estimation of the Surface Free Energy of Polymers. *Journal of Applied Polymer Science 13*, 1741–1747.
10. Siciński, M., Bieliński, D., Gozdek, T., Piątkowska, A., Kleczewska, J. & Kwiatos, K. (2013). Kompozyty Elastomerowe *z* Dodatkiem Grafenu Lub MWCNT Modyfikowanych Plazmochemicznie. *Inżynieria Materiałowa 6,*
11. Xu, T., Yang, J., Liu, J. & Fu, Q. (2007). Surface Modification of Multi-Walled Carbon Nanotubes by O_2 Plasma. *Applied Surface Science, 253*, 8945–8951.

CHAPTER 7

A STUDY ON RADIATION CROSSLINKING OF ACRYLONITRILE-BUTADIENE RUBBER

KATARZYNA BANDZIERZ, DARIUSZ M. BIELINSKI,
ADRIAN KORYCKI, and GRAZYNA PRZYBYTNIAK

CONTENTS

ABSTRACT

Radiation crosslinking of elastomers has been receiving increasing attention. The reactions induced by high-energy ionizing radiation are very complicated and the mechanisms still remain not entirely comprehended. Ionizing radiation crosslinking of acrylonitrile-butadiene rubber, filled with 40 phr of silica, with incorporated sulfur crosslinking system was the object of study. To investigate the influence of components such as sulfur and crosslinking accelerator-dibenzothiazole disulfide (DM) on the process, a set of rubber samples with various sulfur to crosslinking accelerator ratio was prepared and irradiated with 50, 122 and 198 kGy. Crosslink density and crosslink structure were analyzed and mechanical properties of the rubber samples were determined. Inhibiting effect of DM and sulfur on the radiation crosslinking process was found. The rubber vulcanizates having sulfur in composition characterized themselves with hybrid crosslinks–both carbon-carbon and sulfide, with varying degree of sulphidity, depending on the sulfur and DM content. The presence of sulfide crosslinks increased tensile strength of the rubber samples.

7.1   INTRODUCTION

Radiation modification of polymer materials has been gaining increasing popularity, not only in academic research, but also in industrial applications [1]. Among numerous advantages of radiation modification method, the noteworthy issue is the simplicity to control the ionizing radiation dose, which is absorbed by the modified material, dose rate and energy of ionizing radiation. The resulting properties can be therefore 'tailored' and the whole process is highly controllable and repeatable.

Radiation crosslinking is an interesting alternative for thermal crosslinking [2–4] or its complement [5–7]. One of extensively studied polymers in respect to its radiation crosslinking, is acrylonitrile-butadiene rubber (NBR) [8–12], which belongs to group of polymers, which effectively crosslink on irradiation with ionizing radiation.

As a result of high-energetic irradiation, radicals are generated directly on polymer chains. By recombination, they form carbon-carbon (C-C) crosslinks between the chains. Due to the fact that radiation crosslinking leads to formation of C-C crosslinks and the mechanism is radical, it is often compared to peroxide crosslinking [3, 13]. It is noteworthy to enhance that the processes induced by ionizing radiation are very complicated and therefore not thoroughly understood [14]. C-C crosslinks provide good elastic properties and are resistant to thermal aging, but they are short, stiff and are do not provide satisfactory properties for dynamic loadings.

According to Dogadkin's theory, crosslinks of various structure provide better mechanical properties, owing to different lengths of the bridges between polymer chains, which do not break at the same time [15]. To provide optimal properties required for the end-use product, the researchers endeavor to design materials with hybrid (mixed) crosslinks. To obtain hybrid type network with both C-C and longer sulfide crosslinks, studies on thermal simultaneous crosslinking with two types of curatives, such as organic peroxide and sulfur crosslinking system, were carried out. The results generally showed lowered efficiency of crosslinking due to competing reactions, in which sulfur and crosslinking accelerator are involved in reactions with peroxide radicals [16–19].

In our previous research [20] concerning radiation crosslinking of NBR rubber with sulfur crosslinking system in composition, hybrid crosslink structure was proved to be formed upon irradiation. Inhibiting effect of sulfur crosslinking system on total crosslink density, formed in the irradiation process, was observed. To investigate in detail the contribution coming from particular components of the crosslinking system, such as rhombic sulfur and crosslinking accelerator DM, a set of samples with various ratios of sulfur and accelerator was prepared. The effect of these two components on radiation crosslinking process was studied by determination of basic mechanical properties, total crosslink density and analysis of crosslink structure.

7.2 EXPERIMENTAL PART

7.2.1 MATERIALS

Acrylonitrile–butadiene rubber Europrene N3325 (bound ACN content 33%) was supplied by Polimeri Europa (Italy). Precipitated silica Ultrasil VN3 was obtained from Evonik Industries (Germany). Vinyltrimethoxysilane U-611 was obtained from Unisil (Poland). Rhombic sulfur was provided by Siarkopol Tarnobrzeg (Poland) and zinc oxide, stearic acid and dibenzothiazole disulfide (DM), by Lanxess (Germany).

TABLE 7.1   Composition of Rubber Mixes. The Samples are Designated as x/y. Here, x Indicates the Amount of Sulfur and y Amount of Dibenzothiazole Disulfide, Respectively

Rubber mixes x/y Component, wt.%	0/0	0/1.5	2/0	2/1.5
NBR, Europrene N3325	100	100	100	100
Silica, Ultrasil VN3	40	40	40	40
Silane, U611	4	4	4	4
Zinc oxide, ZnO	5	5	5	5
Stearic acid	1	1	1	1
Rhombic sulfur, S_8	0	0	2	2
Dibenzothiazole disulfide, DM	0	1.5	0	1.5

7.2.2   SAMPLES PREPARATION

Rubber mixes were prepared in two-stage procedure. In the first stage, rubber premixes of NBR, filled with 40 phr of precipitated silica and vinyltrimethoxysilane in amount of 10 wt.% of silica in the composite mix, were prepared with the use of Brabender Plasticorder internal micromixer (Germany) at temperature of mixing chamber of 120 °C, with rotors speed of 20 RPM during components incorporation and 60 RPM during 25 min lasting homogenization process. In the second stage, components of crosslinking system, such as zinc oxide, stearic acid, rhombic sulfur and dibenzothiazole disulfide (DM), were incorporated into the premix with David Bridge two-roll open mixing mill (UK) at 40°C and homogenized for 10 min. The samples composition is given in Table 7.1.

   1 mm rubber mixes sheets were compression molded in an electrically heated press at temperature of 110°C under pressure of 150 bar for 4 min.

7.2.3   SAMPLES IRRADIATION

The molded rubber sheets were subjected to electron beam (EB) irradiation at Elektronika 10/10 linear electron accelerator (Russia), located at the Institute of Nuclear Chemistry and Technology (Poland). The absorbed doses were 50, 122 and 198 kGy. Irradiation process was carried out in air atmosphere at room temperature. The rubber sheets were placed horizontally in the front of pulsed, scanned beam. The total doses were obtained by multipass exposure (approx. 25 kGy per pass).

7.2.4 SAMPLES CHARACTERIZATION

7.2.4.1 CROSSLINK DENSITY DETERMINATION

Total crosslink density of the irradiated samples was determined taking advantage of equilibrium swelling in toluene and calculated on the basis of Flory-Rehner equation [21]. The Flory-Huggins interaction parameter used in the calculations for toluene–NBR rubber was 0.435 [22].

7.2.4.2 CROSSLINK STRUCTURE DETERMINATION

The crosslink structure was analyzed and quantified by thiol–amine analysis, which is based on treatment of the crosslinked material with a set of thiol–amine chemical probes, specifically cleaving particular crosslinks types [23]. Polysulfide crosslinks are cleaved by treatment of crosslinked rubber samples with 2-propanethiol (0.4 M) and piperidine (0.4 M) in toluene for 2 h under inert gas atmosphere (argon) at room temperature, while polysulfide and disulfide crosslinks can be cleaved by treatment under the same conditions with 1-dodecanethiol (1 M) in piperidine for 72 h.

7.2.4.3 MECHANICAL PROPERTIES

Mechanical tests were carried out with the use of 'Zwick 1435' universal mechanical testing machine (Germany), according to ISO 37. The crosshead speed was 500 mm/min and the temperature was 23±2 °C. Five dumbbell specimens were tested for each sample and the average is reported here.

7.3 RESULTS AND DISCUSSION

7.3.1 CROSSLINK DENSITY

The crosslink densities of samples irradiated with doses of 50, 122 and 198 kGy, calculated from equilibrium swelling in toluene, are presented in Fig. 7.1. For all samples studied, crosslink densities formed during EB irradiation process are increasing linear function of dose.

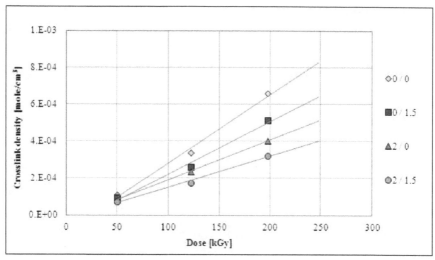

FIGURE 7.1   Total crosslink density as a function of ionizing radiation dose.

The inhibiting effect of DM and sulfur on the radiation crosslinking process was observed. According to experimental work, the inhibiting effect of sulfur (sample 2/0) and DM (sample 0/1.5) is not additive, comparing to corresponding inhibition coming from the same amount of sulfur and DM combined in one sample (2/1.5). The "experimental inhibition" (sample 2/1.5) is lower that the "theoretical inhibition" (summed up inhibition of samples 2/0 and 0/1.5), as shown in Fig. 7.2. The probable explanation of the fact can be sulfur–accelerator complex formed during sheets molding process at 110°C. The complex formed facilitate formation of sulfide crosslinks and possibly makes the reactions more effective–sulfur is used rather for formation of bridges between the polymer chains, than for formation of cyclic structures modifying the chains.

Inhibiting effect of DM arises from the presence of aromatic rings in its structure. The aromatic compounds are known to influence the radiation induced modification by effect of resonance energy dissipation [24–25]. In the structural formula of DM, heteroatoms, such as sulfur and nitrogen are also present. The sulfur moieties are known to inhibit the effect of ionizing radiation action on matter [26–27], due to the fact that sulfur groups act as sinks of the radiation energy [28].

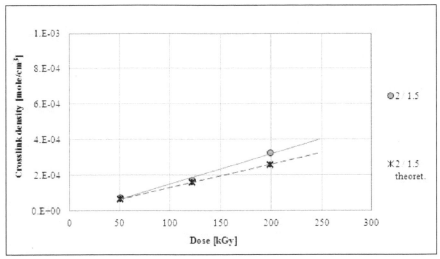

FIGURE 7.2   "Experimental inhibition" (solid line) and "theoretical inhibition" (dashed line) of radiation crosslinking process.

Rhombic sulfur itself also causes large inhibiting effect of subjected to ionizing radiation polymer. It has to be enhanced, that sulfur undoubtlessly is strong radiation-protecting agent, but probably the crosslinking efficiency is reduced because of intramolecular reactions, which result in modification of polymer chains by sulfur cyclic structures. The observed inhibiting effect of sulfur has therefore twofold contribution.

7.3.2 CROSSLINK STRUCTURE

For samples with rhombic sulfur in composition, the crosslink structure investigation was carried out (Fig. 7.3). Presence of both C-C and polysulfide crosslinks was proved. The C-C crosslinks are "regular" effect of polymer irradiation. The sulfide crosslinks were formed as a result of breakage of S-S bonds in highly puckered ring structure of rhombic sulfur by the action of accelerated electrons. The S-S bond energy is low–approx. 240 kJ/mol, what makes it susceptible to break and generate sulfur radicals [6], what consequently leads to sulfide crosslinks formation.

The crosslink structure study showed that during irradiation of the sample containing sulfur, but without crosslinking accelerator (sample 2/0), the participation of polysulfide crosslinks in the total crosslink density is approx. 40%. The difference between the number of polysulfide crosslinks formed upon irradiation with 122 and 198 kGy is very little. In sample 2/1.5 in which both sulfur and crosslinking accelerator are present, the number of polysulfide crosslinks is lower than in sample 2/0, and it slightly increases with irradiation dose (from 28% for 122 kGy up to 32% for 198 kGy). The presence of complex of crosslinking accelerator with sulfur promoted thereby formation of shorter crosslinks.

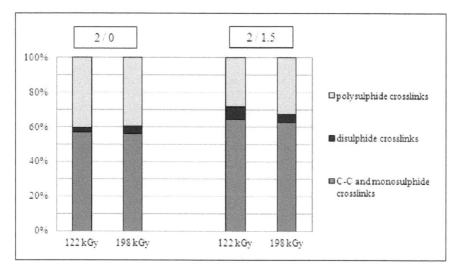

FIGURE 7.3   Crosslink structure of samples 2/0 and 2/1.5, irradiated with 122 and 198 kGy. Network density formed in the samples irradiated with a dose of 50 kGy was very low and the results on crosslink structure obtained from the thiol–amine analysis was not reliable.

7.3.3   MECHANICAL PROPERTIES

The mechanical properties of all samples studied are presented in Table 7.2.

TABLE 7.2 Mechanical Properties (SE_{100}, SE_{200}, SE_{300}, TS, E_b) of Samples Irradiated with 50, 122 and 198 kGy

Sample	Dose [kGy]	Crosslink density [mol/cm³]	Mechanical properties				
			SE_{100} [MPa]	SE_{200} [MPa]	SE_{300} [MPa]	TS [MPa]	E_b [MPa]
0/0	50	1.1×10^{-4}	2.7	4.0	5.6	10.6	599
	122	3.4×10^{-4}	5.1	10.2	17.2	25.1	397
	198	6.6×10^{-4}	9.8	21.1	–	23.6	219
0/1.5	50	9.3×10^{-5}	2.0	2.9	4.0	7.6	670
	122	2.6×10^{-4}	3.9	7.6	12.5	20.8	441
	198	5.1×10^{-4}	7.5	15.8	–	23.7	277
2/0	50	8.3×10^{-5}	1.9	2.6	3.5	7.6	756
	122	2.4×10^{-4}	4.2	7.8	12.5	25.2	503
	198	4.0×10^{-4}	6.1	12.6	21.6	30.9	388
2/1.5	50	7.3×10^{-5}	2.1	2.7	3.5	6.5	756
	122	1.7×10^{-4}	3.0	5.3	8.2	17.4	549
	198	3.2×10^{-4}	4.8	9.7	16.3	26.7	429

In sample 2/0, the generated sulfur radicals inserted into polymer chains, forming long, polysulfide crosslinks, which have significant participation in the total crosslink density. The presence of polysulfide crosslinks is evident in mechanical properties–high tensile strength is provided by these long, labile bridges, which effectively dissipate the energy. Due to this effect, sample 2/0 showed the highest tensile strength among all analyzed samples (Fig. 7.4). The lowest value of tensile strength exhibited the sample 0/1.5. The presence of the DM not only inhibited the formation of crosslinks, but also considerably deteriorated the resulting mechanical properties of the rubber sample. Modification of the polymer chain by the products of DM transformation upon irradiation is probable.

Tensile strength curve of sample containing both sulfur and DM, is located between the corresponding curves of samples containing solely DM or sulfur. Its tensile strength is higher than of sample without sulfur nor DM (0/0), due to presence of mixed, diversified crosslinks in sample 2/1.5, and exclusively uniform C-C crosslinks in sample 0/0.

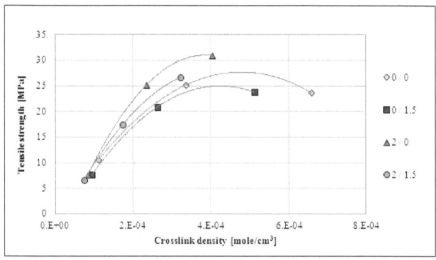

FIGURE 7.4   Tensile strength of samples as a function of crosslink density.

7.4   CONCLUSIONS

In our study, the influence of particular components of sulfur crosslinking system, such as rhombic sulfur and crosslinking accelerator DM, on the process of radiation crosslinking of NBR was investigated.

1. Inhibition of radiation crosslinking by both sulfur and DM was proved. Due to complex nature of the investigated system and complicated processes induced by high-energy radiation, it is difficult to unambiguously identify the mechanisms responsible for the inhibiting effect.

2. Irradiation of samples with sulfur in composition leads to formation of hybrid network type, characterizing itself with both short C-C crosslinks and longer sulfide ones. The presence of diversified crosslinks guarantee high tensile strength of the rubber samples.

3. The reactions induced in polymer matrix with sulfur crosslinking system in composition are probably multistage. To comprehend mechanisms of the reactions initiated by ionizing radiation, further investigation within this area is needed.

ACKNOWLEDGMENTS

The work was performed in the frame of Young Scientists' Fund at the Faculty of Chemistry, Lodz University of Technology, Grant W-3/FMN/6G/2013.

KEYWORDS

- Crosslink density and structure
- Mechanical properties
- Nitrile rubber
- Radiation crosslinking

REFERENCES

1. Clough, R. L. (2001). High-Energy Radiation and Polymers: A Review of Commercial Processes and Emerging Applications. *Nucl. Instrum. Meth. B.*, *185(1–4)*, 8–33.
2. Bhowmick, A. K. & Vijayabaskar, V. (2006). Electron Beam Curing of Elastomers. Rubber *Chem. Technol. 79(3)*, 402–428.
3. Manaila, E., Stelescu, M. D. & Craciun, G. (2012). Advanced Elastomers–Technology, Properties and Applications. Aspects Regarding Radiation Crosslinking of Elastomers. In-Tech, 3–34.
4. Bik, J., Gluszewski, W., Rzymski, W. M. & Zagorski Z. P. (2003). EB radiation crosslinking of elastomers. *Radiat. Phys. Chem. 67(3)*, 421–423.
5. Stepkowska, A., Bielinski, D. M. & Przybytniak, G. (2011). Application of Electron Beam Radiation to Modify Crosslink Structure in Rubber Vulcanizates and Its Tribological Consequences, *Acta Phys. Pol. A.*, *120(1)*, 53–55.
6. Vijayabaskar, V., Costa, F. R. & Bhowmick, A. K. (2004). Influence of Electron Beam Irradiation as one of the Mixed Crosslinking Systems on the Structure and Properties of Nitrile Rubber. *Rubber Chem. Technol. 77(4)*, 624–645.
7. Vijayabaskar, V. & Bhowmick, A. K. (2005). Dynamic Mechanical Analysis of Electron Beam Irradiated Sulfur Vulcanized Nitrile Rubber Network—Some Unique Features. *J. Mater. Sci.*, *40(11)*, 2823–2831.
8. Yasin, T., Ahmed, S., Yoshii, F. & Makuuchi, K. (2002). Radiation Vulcanization of Acrylonitrile-Butadiene Rubber with Polyfunctional Monomers. *React. Funct. Polym.*, *53(2–3)*, 173–181.
9. Bik, J. M., Rzymski, W. M., Gluszewski, W. & Zagorski, Z. P. (2004). Electron Beam Crosslinking of Hydrogenated Acrylonitrile-Butadiene Rubber. *Kaut. Gummi Kunstst. 57(12)*, 651–655.
10. Stephan, M., Vijayabaskar, V., Kalaivani, S., Volke, S., Heinrich, G., Dorschner, H., Wagenknecht, U. & Bhowmick, A. K. (2007). Crosslinking of Nitrile Rubber by Electron Beam Irradiation at Elevated Temperatures. *Kaut. Gummi Kunstst.*, *60(10)*, 542–547.
11. Vijayabaskar, V., Tikku, V. K., & Bhowmick, A. K. (2006). Electron Beam Modification and Crosslinking: Influence of Nitrile and Carboxyl Contents and Level of Unsaturation on Structure and Properties of Nitrile Rubber. *Radiat. Phys. Chem.*, *75(7)*, 779–792.

12. Hill, D. J. T., O'Donnell, J. H., Perera, M. C. S., & Pomery, P. J. (1996). An Investigation of Radiation-Induced Structural Changes in Nitrile Rubber. *J. Polym. Sci. Pol. Chem., 34(12)*, 2439–2454.

13. Loan, L. D. (1972). Peroxide Crosslinking Reactions of Polymers. *Pure Appl. Chem., 30(1–2)*, 173–180.

14. Zagorski, Z. P. (2002). Modification, Degradation and Stabilization of Polymers in View of the Classification of Radiation Spurs. *Radiat. Phys. Chem., 63(1)*, 9–19.

15. Dogadkin, B. A., Tarasova, Z. N., Golberg I. I. & Kuanyshev K. G. (1962). Effect of Vulcanization Structures on the Strength of Vulcanizates. *Kolloid. Zh., 24*, 141–151.

16. Manik, S. P. & Banerjee, S. (1969). Studies on Dicumylperoxide Vulcanization of Natural Rubber in Presence of Sulfur and Accelerators. *Rubber Chem. Technol., 42(3)*, 744–758.

17. Manik, S. P. & Banerjee, S. (1970). Sulfenamide Accelerated Sulfur Vulcanization of Natural Rubber in Presence and Absence of Dicumyl Peroxide. *Rubber Chem. Technol., 43(6)*, 1311–1326.

18. Bakule, R. & Havránek, A. (1975). The Dependence of Dielectric Properties on Crosslinking Density of Rubbers. *J. Polym. Sci. Polym. Symp., 53(1)*, 347–356.

19. Bakule, B., Honskus, J., Nedbal, J. & Zinburg, P. (1973). Vulcanization of Natural Rubber by Dicumyl Peroxide in the Presence of Sulfur. *Collect. Czech. Chem. Commun. 38(2)*, 408–416.

20. Bandzierz, K. & Bielinski, D. M. (2013). Radiation Methods of Polymers Modification: Hybrid Crosslinking of Butadiene–Acrylonitrile Rubber. 244–247. Abstracts Collection on New Challenges in the European Area: Young Scientists, Baku, Azerbaijan.

21. Flory, P. J. & Rehner, J. (1943). Statistical Mechanics of Crosslinked Polymer Networks II. Swelling. *J. Chem. Phys., 11(11)*, 521–526.

22. Hwang, W.-G., Wei, K.-H. & Wu, C.-M. (2004). Mechanical, Thermal, and Barrier Properties of NBR/Organosilicate Nanocomposites. *Polym. Eng. Sci. 44(11)*, 2117–2124.

23. Saville, B. & Watson, A. A. (1967). Structural Characterization of Sulfur-Vulcanized Rubber Networks. *Rubber Chem. Technol., 40(1)*, 100–148.

24. Głuszewski, W. & Zagórski, Z. P. (2008). Radiation Effects in Polypropylene/Polystyrene Blends as the Model of Aromatic Protection Effects. *Nukleonika. 53(1)*, 21–24.

25. Seguchi, T., Tamura, K., Shimada, A., Sugimoto, M. & Kudoh, H. (2012). Mechanism of Antioxidant Interaction on Polymer Oxidation by Thermal and Radiation Ageing. *Radiat. Phys. Chem. 81(11)*, 1747–1751.

26. Charlesby, A., Garratt, P. G. & Kopp, P. M. (1962). Radiation Protection with Sulfur and Some Sulfur-Containing Compounds. *Nature. 194*, 782.

27. Charlesby, A., Garratt, P. G. & Kopp, P. M. (1962). The Use of Sulfur as a Protecting Agent Against Ionizing Radiations. *Int. J. Radiat. Biol., 5(5)*, 439–446.

28. Nagata, C. & Yamaguchi, T. (1978). Electronic Structure of Sulfur Compounds and Their Protecting Action Against Ionizing Radiation. *Radiat. Res., 73(3)*, 430–439.

CHAPTER 8

A CASE STUDY ON SORPTION PROPERTIES OF BIODEGRADABLE POLYMER MATERIALS

MARINA BAZUNOVA, IVAN KRUPENYA, ELENA KULISH, and GENNADY E. ZAIKOV

CONTENTS

ABSTRACT

Compositions obtained ultrafine powders based on low-density polyethylene, a modified natural polymer chitosan under the combined influence of high-pressures and shear deformation. Studied their sorption properties and susceptibility to biodegradation. The resulting samples of polymer films based on low density polyethylene modified chitosan having acceptable strength characteristics, good absorbent capacity and biodegradability, can be used for manufacturing biodegradable packaging materials.

8.1   INTRODUCTION

For materials for the manufacture of food packaging, disposable products, it is advisable the use of biodegradable polymers, which retain only the performance during the period of consumption, and then undergo a physicochemical and biological transformations under the influence of environmental factors, and easily incorporated into natural metabolic processes bio systems.

The problem of biodegradability of well-known tonnage industrial polymers is quite urgent for modern studies. It is promising enough to use synthetic and natural polymer mixtures which can play the roles of both filler and modifier for creating biodegradable environmentally safe polymer materials. The macromolecule fragmentation of the synthetic polymer is to be provided for due to its own biodestruction.

The synthetic polymers have been modified by the natural one under the combined effect of high pressure and shear deformation. The usage of this method for obtaining polymer composites is sure to solve several problems at once. Firstly, the ultra dispersed powders with a high homogeneity degree of the components can be obtained under combined high pressure and shear deformation thus resulting in easing the technological process of production [1]. Secondly, the elastic deformation effects on the polymer material may lead to the chemical modification of the synthetic polymer macromolecules by the natural polymer blocks via recombination of the formed radicals. Thus, it can provide for the polymer product biodegradation.

As components in the preparation of biodegradable polymeric composite synthetic polymer used a large-capacity, low density polyethylene (LDPE) and naturally occurring polysaccharide chitosan (CTZ).

The parameters characterizing the tendency of the compositions to biodegradation selected by their ability to absorb water (degree of swelling) and the mass loss by maintaining samples in soil. Water absorption is one of the indirect indicators of the propensity of the material to biodegradation, as the swollen material accelerated diffusion processes synthesized by microorganisms enzymes catalyzing the process of biological degradation [2].

Therefore, is appropriate to study the sorption properties and ready biodegradability compositions based on ultrafine powders LDPE modified natural polymers HTZ under the combined effects of high-pressure and shear deformation.

8.2   EXPERIMENTAL PART

LDPE 10803-020 (90,000 molecular weight, 53% crystallinity degree, and 0,917 g/sm^3 density) and chitosan samples of Bioprogress Ltd. (Russia) obtained by alkalinedeacetylation of crab chitin (deacetylation degree ~84%), and $M_{sd} = 115000$ were used as components for producing biodegradable polymer films.

The initial highly dispersed powders with different mass ratio of components have been obtained by high temperature shearing (HTS) under simultaneous impact of high pressure and shear deformation in an extrusion type apparatus with a screw diameter of 32 mm [3, 4]. Temperatures in kneading, compression and dispersion chambers amounted to 150°C, 150°C and 70°C, respectively.

The size of particles in powders of LDPE, CTZ and LDPE/CTZ with various mass ratio of the components were determined by "Shimadzu Salid–7101" particle size analyzer. The film formation was carried out by rotomolding [5] at 135 and 150°C. The film sample thickness amounted to 100 um and 800 um.

As a measure of the degree of modification (P, HTZ grams to 1 gram polyethylene) adopted mass HTZ who "grafted" onto polyethylene. Here it is assumed that HTZ chemically bound to polyethylene, insoluble in 1% acetic acid solution, unlike the unbound HTZ.

The absorption coefficient of the condensed vapors of volatile liquid (water, n-heptane) K in static conditions is determined by complete saturation of the sorbent by the adsorbent vapors under standard conditions at 20°C [6] and was calculated by the formula: $K' = \dfrac{m_{absorbed\ water}}{m_{sample}} \times 100\%$, where m$_{absorbed\ water}$ is weight of the saturated condensed vapors of volatile liquid, g; m$_{sample}$ is weight of dry sample, g.

Film samples were long kept in the aqueous and enzyme media to determine the water absorption coefficient while the absorbed water weight was calculated. The water absorption coefficient of film samples of LDPE/CTZ with different weight ratio was determined by the formula: $K = \dfrac{m_{absorbed\ water}}{m_{sample}} \times 100\%$

, where $m_{absorbed\ water}$ is water weight absorbed by the sample whereas m_{sample} is the sample weight. Sodium azide was added to the enzyme solution to prevent microbial contamination. Each three days both the water medium and the enzyme solution were changed. The "Liraza" agent of 0.1 g/L concentration was used as an enzyme (Immunopreparat SUE, Ufa. Russia). In experiments for determining the absorption of the condensed vapors of volatile liquid and water absorption coefficients at a confidence level of 0,95 and 5 repeated experiments, the error does not exceed 7%.

In assessing the activity of composites to bind the protein as a biological marker albumin solution used in the model [7] obtained by precipitating casein from nonpasteurized milk, followed by separating the casein by centrifugation. The concentration of albumin to sorption and thereafter determined spectrophotometrically according to the formula:

Protein content = $1.45 \times D_{280} - 0.74 \times D_{260}$ (mg/mL)

where D_{280} – absorbance of the solution at 280 nm; D_{260} – absorbance of the solution at 260 nm.

The obtained film samples were kept in soil according to the method [8] to estimate the ability to biodegradation. The soil humidity was supported on 50–60% level. The control of the soil humidity was carried out by the hygrometer ETR-310. Acidity of the soil used was close to the neutral with pH = 5.6–6.2 (pH-meter control of 3in1 pH). At a confidence level 0.95 and 5 repeated experiments the experiment error in determining the tensile strength and elongation does not exceed 5%.

Mechanical film properties (tensile strength (σ) and elongation (ε)) were estimated by the tensile testing machine ZWIC Z 005 at 50-mm/min tensile speed.

8.3   RESULTS AND DISCUSSION

It can be assumed that the process of elastic-strain effects on the mixture of LDPE and CTZ can lead to chemical modification of macromolecules synthetic polymer blocks natural polymer due to the recombination of macroradicals formed. In this regard, the evaluated degree of modification of polyethylene by treating chitosan superfine powders LDPE/CTZ obtained by HTS, an excess of 1%-solution of acetic acid. When it is allowed that chitosan chemically linked polyethylene, is insoluble in 1% acetic acid solution, unlike the unbound chitosan. Data on the extent of modification of polyethylene chitosan (P) and on the proportion of (D) CTZ, which entered the process in the modification with polyethylene on its total weight, are shown in Table 8.1.

From Table 8.1, it follows that the combination of the initial components under HTS leads to a fairly high degree of modification of chitosan macro-molecules polyethylene fragments.

The speed of the hydrolytic destruction of the polymer materials is closely connected with their ability to water absorption. Values of their absorption capacity according to water and heptane vapors were determined for a number of powder mixture samples of LDPE/CTZ (Table 8.2). It was established that the absorption coefficient of the condensed water vapors is directly propor-tional to the chitosan content.

As the initial powders, the films with high chitosan content under roto-molding absorb water well (Table 8.3). At the same time thinner films absorb more water for a shorter period of time.

TABLE 8.1 The Degree of Modification of Polyethylene Chitosan (P) and Share (D) HTZ, Entered into the Process of Modifying Polyethylene During HTS/

№	LDPE/CTZ powder, mass.%	Particle size, um	P, g HTZ/1 g LDPE	D, %
1	0	5.5–8.0; 10.0–80.0	–	–
2	20	6.5–63.0	0.19	25.7
3	40	6.5–50.0	0.32	38.3
4	50	4.3–63.0	0.45	44.6
5	60	6.5–63.0	0.73	48.4

TABLE 8.2 The Absorption Coefficient of the Condensed Water Vapors of Volatile Liquid (water and n-heptane) $\overset{\cdot}{K}$ of LDPE/CTZ Powders at 20°C.

№	LDPE/CTZ powder, mass.%	$\overset{\cdot}{K}$ by water vapors, %	$\overset{\cdot}{K}$ by n-heptane, %
1	0	1.10±0.08	17±1.0
2	20	12.3±0.8	11.0±0.8
3	40	20±1.0	5.0±0.4
4	50	25±2.0	4.0±0.3
5	60	35±2.0	4.0±0.3

TABLE 8.3   Values of Equilibrium Water Absorption Coefficients K (%) of LDPE/CTZ Films at 20°C.

№	LDPE/CTZ powder, mass.%	K, %			
		Medium–water		Medium–Liraza enzyme (0.1 g/L)	
		Film thickness 100 um	Film thickness 800 um	Film thickness 100 um	Film thickness 800 um
1	20	5.0±0.4	2.0±0.2	5.0±0.4	4.0±0.3
2	40	10.0±0.7	4.0±0.3	13.0±0.9	7.0±0.5
3	50	38±3	14±1.0	40±3.0	45±3.0
4	60	-	31±2.0	-	95.8±0.7

In case the film samples were placed into the enzyme solution, water absorption changes slightly. Firstly, the equilibrium values of the absorption coefficient of films in the enzymatic medium are higher than in water (Table 8.3). It is in the enzymatic medium usage that a longer film exposure (for more than 30–40 days) was accompanied by weight losses of the film samples. Moreover, after 40 days of testing, the film with 50%mass of chitosan and 100um thickness lost its integrity. Films of 800 um thick and chitosan content of 50 and 60% lost their integrity after 2 months of the enzyme agent solution contact (Fig. 8.1). These facts are quite logical as "Liraza" is subjected to a β-glycoside bond break in chitosan. Thus, the destruction of film integrity is caused by the biodestruction process. Higher values of the water absorption coefficient may be explained by enzyme destruction of chitosan chains as well due to some loosening in the film material structure (Table 8.3).

In assessing the activity of composites to bind the protein as a biological marker albumin solution used in the model. Found that the maximum of activity to bind a protein after exposure of samples in the model solution for 24 h have LDPE film/HTZ composition 40/60 wt.% Lowering serum albumin by about 17%. Consequently, the composites obtained showing sorption activity in relation to the nature of biological markers.

Tests on holding the samples in soil indicate on bio destruction of the obtained film samples either. It is found that the film weight is reduced by 7–8% during the first six months. Here the biggest weight losses are observed in samples with 50–60 mass% of chitosan.

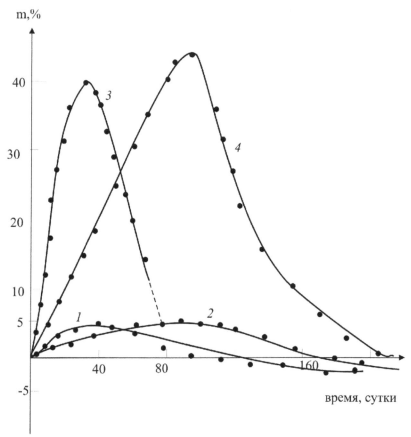

FIGURE 8.1 Curves of relative mass change of the film samples LDPE/ CTZ, immersed in a solution of the enzyme preparation "Liraza" concentration of 0.1 g/l (20 °C): (1)-CTZ content of 20 wt.%, the film thickness of 100 um, (2)-the contents CTZ 20 wt.%, the film thickness of 800 um, (3)-the contents CTZ 50 wt.%, the film thickness of 100 um, and (4)-the contents CTZ 50 wt.%, the film thickness of 800 um/

Chitosan introduction into the polyethylene matrix is accompanied by changes in the physical and mechanical properties of the film materials. The polysaccharide introduction into the LDPE compounds results in slight decrease in the tensile strength of films. Wherein the number of the chitosan introduced does not affect the composition strength. However, low-density polyethylene/chitosan films obtain much less elongation values as compared with low density polyethylene films under the same conditions. Thus, films which were obtained on the basis of ultra dispersed LDPE powders modified by chitosan possess less plasticity while retain in got her satisfactory strength properties.

Thus, are obtained composition based on the ultrafine powders LDPE modified HTZ under the combined influence of high pressures and shear deformation. Samples of polymer films based on LDPE modified HTZ having acceptable strength characteristics, good absorbent capacity and biodegradability, can be used for manufacturing biodegradable packaging materials.

KEYWORDS

- **Biodegradable polymer films**
- **Chitosan**
- **Low-density polyethylene**
- **Water-absorbent capability**

REFERENCES

1. Bazunova, M. V., Babaev, M. S., Bildanova, R. F., Protchukhan, Yu.A., Kolesov, S. V. & Akhmetkhanov, R. M. (2011). Powder-Polymer Technologies in Sorption-Active Composite Materials. *Vestn. Bashkirs. Univer., 16(3), 684–688.*
2. Fomin, V. A. & Guzeev, V. V. (2001). Biodegradable Polymers, Condition and Prospects. *Plasticheskie Massi, 2,* 42–46.
3. Enikolopyan, N. S., Fridman, M. L. & Karmilov, A. Yu. (1987). Elastic-Deformation Grinding of Thermo-Plastic Polymers. *Reports AS USSR, 296(1),* 134–138.
4. Akhmetkhanov, R. M., Minsker, K. S. & Zaikov, G. E. (2006). On the Mechanism of Fine Dispersion of Polymer Products at Elastic Deformation Effects. *Plasticheskie Massi, 8,* 6–9.
5. Sheryshev, M. A. (1989). *Formation of Polymer Sheets and Films.* Ed. Braginsky V.A. Leningrad: Chemistry Publishing, 120 p.
6. Keltsev, N. V. (1984). *Fundamentals of Adsorption Technology.* Moscow: Chemistry, 595 p.
7. Asher, W. J., Davis, T. A. & Klein, E. (1989). *Sorbents and Their Clinical Application.* ed. C. Giordano. Kiev: Naukova Dumka, 398 p.
8. Ermolovitch, O. A., Makarevitch, A. V., Goncharova, E. P. & Vlasova, F. M. (2005). Estimation Methods of Biodegradation of Polymer Materials. *Biotechnology, 4,* 47–54.

CHAPTER 9

THE INFLUENCE OF POLYAMIDE MEMBRANES ON ULTRAFILTRATION PROCESS PRODUCTIVITY

E. M. KUVARDINA, F. F. NIYAZY, R. Y. DEBERDEEV, G. E. ZAIKOV, and N. V. KUVARDIN

CONTENTS

ABSTRACT

This chapter concerns questions of division of multicomponent solutions by means of polyamide membranes in the course of ultrafiltration. The question of influence of low-frequency fluctuations on a polyamide membrane for the purpose of increase of its productivity is considered.

9.1   INTRODUCTION

Pellicular semipermeable membranes that are used for separation of multi-component solutions, have a two-layer structure.

Polymer membranes are formed primarily by methods based on solution processing of polymers (less frequently–on polymer melts). There are various methods for membranes production: irrigating out of solution, molding from the melt, forming pores in foils by nuclear particles, followed by leaching of the degradation products, pressing, rolling, etc.; As a result, membranes have a system of through pores, which form the labyrinths of interconnected channels, the diameter of pores by depth of membrane can be changed due to their isotropy.

According to thickness such membranes are unequal: the surface layer is responsible for the selectivity of separation, denser layer (its thickness is 20–40 µm), an inner layer, which is more coarse-pored (the thickness is 100–200 µm) is used as the substrate and improves the mechanical strength of membrane, also have a big impact on the formation of selective blocking layer.

Operating layer is made mainly of synthetic polymers, copolymers, and their mixtures. The most popular practical application during the ultrafiltration is obtained by membranes, working layer of which is based on cellulose and its derivatives (cellulose acetate), and polyamides. [1]

In our work we used the membrane named UPM-50P, working layer of which is made of polysulfonamide.

Physical-and-chemical properties of the membranes depend largely on the nature of the material, which they are formed of, and methods for their preparation. They are defined by chemical structure and composition of the polymer, which determine such important characteristics as the flexibility of the polymer chain, the origin and the energy of intermolecular interactions, as well as interactions with components of separating solutions. Physical-and-chemical and mechanical properties of membranes are largely defined by molecular structure.

The mechanical properties of polymers are: mechanical strength, deformability, ability to develop reversible and irreversible deformation, fatigue resistance in multiple strains, detrition.

In the matter of definition of the most rational membrane unit parameters it is necessary to consider the physical condition of the membranes during their operations in the separation of multicomponent solutions, considering major external factors affecting the working surface of membrane [2]. During operation of membrane unit external factors can be considered as the temperature, values of which increases by direct proportion since labor hours, chemical composition of separating solution, operating pressure and mode of its changing.

9.2 EXPERIMENTAL PART AND PROCESSING OF RESULTS

This chapter describe the influence of temperature and pressure on the motion of mechanical properties of polysulfonamide membrane. As the part of membrane's mechanical characteristics, we consider the value of membrane deformations arising under a vibration of the circulating solution.

It is known that one of the factors that influence the mechanical properties of polymers is temperature.

Melting temperature of polyamides is within the range 65–88 °C. Values of entropy and fusion heat of polyamides calculated theoretically and confirmed experimentally are described by the following equation:

$$\Delta H = T_{melt} \cdot \Delta S; \tag{1}$$

where ΔH is fusion heat [kJ/mol]; T_{melt} is melting point of polyamide [K]; ΔS is entropy change [kJ/(mol K)] [3].

Melting temperature of polysulfonamide is in the range of 65±20 °C.

Physical properties of polysulfonamides, generally, are linearly unaffected to the composition of polysulfonamide. They have a minimum in the curve of the melting temperature of the composition, mostly, in the equimolar ratio of elementary units or nearly equimolar ratio.

Thus, the maintained temperature of ultrafiltration was in the range of 60 °C. Temperature above 60 °C may be unsuitable for the quality of solutions, that is the reason why during our practical work we use temperature range of 4–60 °C.

As a partial solution we use a raw juice of sugar beets. Pressure generate in the machine equals to 0.25–0.3 MPa.

As a result, due to the pressure generated in the machine, membrane is compressed. Moving of mechanical compression energy into heat energy increases the temperature of the membrane [4]. However, its value disproportionately low compared to the temperature of circulating solution, so we do not consider it. The upper temperature limit is 60 °C. We cannot take into account only this temperature limit, because it does not influence on mechanical properties of membrane, inasmuch as it does not lead to a change in the physical-and-chemical properties of membrane material.

The curve in Fig. 9.1 illustrates the deceleration of separation process in the initial solution temperature +10 °C. In the context of concentration polarization theory, we can confirm that due to clogging of large pores, deposition layer is formed on its surface without altering the structural properties of the membrane [5].

The greatest interest during the experiment caused the range of low temperatures between +4 °C and 10 °C. Fig. 9.1, curve 2. As we can see, in the initial period the productivity of membrane increases. That fact can be explained by knowing the structural properties of the polymer, so that the membrane material behaves like a solid substance, the density of molecular packing increases, as well as the density of membrane itself.

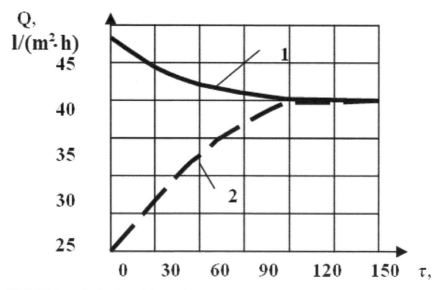

FIGURE 9.1   Indication of the productivity of the membrane during the first time period; 1–initial solution temperature above +10 °C; 2–initial solution temperature +4 °C.

Compressing of the large pores of membrane prevents the formation of swirls [6] so at this stage of the process occurs a phenomenon, when the suspended particles and macromolecular compounds remain on the surface of membrane as colloidal unstable deposits, which are easily washed off by the circulating solution. Low-molecular composition easily enough pass through the small pores of membranes without interrupting. In these circumstances, sorption of gel formation in the pores is virtually eliminated.

At such temperature conditions, deformation of the membrane due to the action of pressure generated into the machine and vibration exposure is inconspicuous.

Increasing the temperature of the solution, membrane starts to behave itself as an elastic substance; otherwise, a change of distances between molecules of polymer membrane layer occurs when periodic action of low-frequency vibration, introduced into circulating solution, creates vibrations of the medium.

Influence of vibration on the mechanical properties of the membrane is investigated in the frequency range between 10 Hz and 1500 Hz [6].

Membrane productivity for all values of the specified frequency range is not changed, remaining within the productivity of the membrane unit, working under pressure without mixing the solution by oscillation and is significantly increased with exposure to low frequency in the range of 60–70 Hz. The greatest productivity into this limit corresponds 67 Hz, Fig. 9.2.

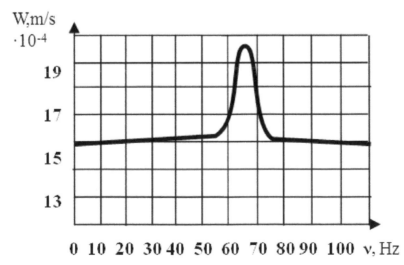

FIGURE 9.2   Dependence of filtration rate on the oscillation frequency introduced into the solution.

Let us try to analyze the effect obtained. It is known that as a criterion for characterizing the viscoelastic properties of the membrane can take a square hysteresis loop, curve described by G=f (P) at gradually increasing pressure from zero to a certain value, and then the pressure change for the reverse sequence. [7]

A typical curve "load – strain" during operation of membrane at a constant pressure is shown in Fig. 9.3. Current size of the hysteresis loop can be represented as a sum of two integrals:

$$S = \int_{0}^{\varepsilon_2} \sigma d\varepsilon + \int_{\varepsilon_2}^{\varepsilon^1} \sigma d\varepsilon \tag{2}$$

where

$$\varepsilon = \frac{(h - h_0)}{h_0}$$

In this case, σ is a function of the relative compression ε, and intersection of '$\sigma d\varepsilon$' under the integral sign has the dimension of work per unit volume

$$\sigma d\varepsilon = \frac{F}{S} \cdot \frac{dh}{h_0} = \frac{Fdh}{v} \tag{3}$$

Thus, the area of the hysteresis loop is the difference between the specific work expended in compression (loading) of the membrane and obtained during unloading.

At entering vibrations into the flow, loading of the membrane takes place unevenly, thus such curves as loading-compression and unloading-straightening do not coincide, and a graph describing these relationships will have a different character as shown in Fig. 9.3.

During unloading the membrane material, we can observe the dependence that same load values have large values of compression and unloading curve misses the origin, indicating the presence of residual strain in the membrane as shown in Fig. 9.4.

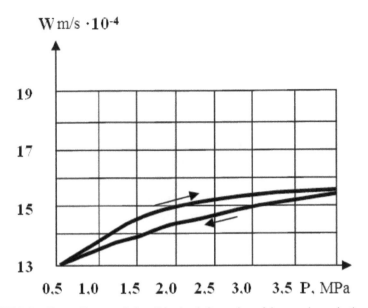

FIGURE 9.3 Shows "hysteresis loop" in the deformation of the membrane in the absence of vibrations introduced into the solution.

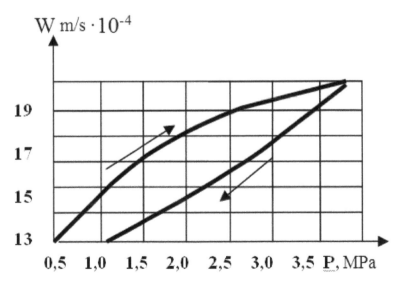

FIGURE 9.4 Shows "hysteresis loop" in the deformation of the membrane in the presence of 60–70 Hz vibrations introduced into the solution.

The hysteresis loop area depends on the rate of force application, in this case, the frequency of the vibrations introduced into the solution. We presume that in the state of the value of vibration frequency below 60 Hz, values of σ and ε are close to equilibrium and the loading and unloading curves are close to each other. This indicates that the area of the hysteresis loop has small value. On the other hand, for values of the vibration frequency above 70 Hz, loading rate of the membrane material is very high and the elements of polymer structure does not have time to regroup, and the material of the membrane is deformed not by kinetic units regrouping, but due to changes in the distances between the particles. In this case, the deformation of the membrane is small, its value when loading and unloading are close to each other, therefore, the area of the hysteresis loop also has a small value.

At temperatures above 10 °C gel layer is formed on the surface of membrane and along with it the sorption takes place with the formation of pores therein additional gel layers.

Vibration frequency of 60–70 Hz helps to prevent gel formation in the pores of the polymer membrane.

9.3   CONCLUSIONS

1.   The surface of hysteresis loop is small as well as at very high speeds as at very low rates of force applying. It has the largest value for such amounts of force processing time, which are comparable with the value of the relaxation time of the membrane.
2.   Within frequency vibrations introduced into the circulating flow, ultrafiltration of sugar beet raw juice corresponds to the values of 60–70 Hz. We can the largest surface of the hysteresis loop, as evidenced by an increasing in productivity of the membrane unit.

KEYWORDS

- Membrane
- Multicomponent solution
- Polyamide
- Process
- Ultrafiltration
- Vibration

REFERENCES

1. Niyazi, F. F., Savenkova, I. V., Zaikov, G. E. & Margolin, A. L. (2013). "Kinetics and Mechanism of Photo-Oxidation of Cellulose Diacetate": Niyazi, F. F., Savenkova, I. V., Zaikov, G. E., Margolin, A. L., Vestnik of the Kazan Technological Institute, *21*, 145–150.
2. Plokhotnikov, S. P., Bogomolova, O. I., Plokhotnikov, D. S., Bogomolov, V. A., Belova, E. N., Harina, M. V. & Nizaev, A. D. (2013). "The Method of Constructing Modified Relative Permeabilities in the Models of Three-Phase Filtering in Layered Reservoirs" [Text]: Plokhotnikov, S. P., Bogomolov, O. I., Plokhotnikov, D. S., Belova, E. H., Harina, M. V. & Nizaev, R. H. Vestnik of the Kazan Technological Institute, *21*, 283–287.
3. Kudryavtsev, G. I., Nosov, M. P. & Volokhina, A. V. (1976). "Polyamide Fibers." [Text]: Kudryavtsev, G. I., Nosov, M. P. & Volokhina, A. V. M.: "Chemistry," 260.
4. Gul, V. E. (1976). "Structure and mechanical properties of polymers." [Text]: Gul, V. E., Kuleznev, V. N. M.: Higher. Sch., 314.
5. Cherkasov, A. N. & Pasechnick, V. A. (1991). "Membranes into Biotechnology and Sorbents." [Text]: Cherkasov, A. N., Pasechnick, V. A., 240.
6. Kuvardina, E. M. (2003). PhD thesis. "Dynamics Ultrafiltration Unit for the Separation of Sugar Beet Diffusion Juice." [Text]: Kuvardina, E. M., Kursk KSTU, 143.
7. Dytnersky, Y. I. (1986). "Boromembrane Processes. Theory and Calculation." [Text]: Dytnersky, Y. I. M.: "Chemistry," 272.

CHAPTER 10

A STUDY ON THERMO-CHEMICAL CHANGES OF NATURAL RUBBER

I. A. MIKHAYLOV, YU. O. ANDRIASYAN, R. JOZWIK,
G. E. ZAIKOV, and A. A. POPOV

CONTENTS

ABSTRACT

Thermo-mechano-chemical changes of natural rubber SVR 3L under treatment internal mixer at self heating have been studied. Effect of molecular mass and content of gel-fraction of natural rubber is shown. Properties of rubber compounds and vulcanized rubber are presented.

10.1   AIMS AND BACKGROUND

Now the world elastomeric market consists of the natural rubber (NR)–40% and synthetic rubbers–60%. According to forecasts of experts the tendency of increase in a share of NR is observed. It is supposed that by 2015–2020 its share will make 50%. For the purpose of expansion of a scope of natural rubber manufacturing of NR with the content of chlorine from 0.5 to 15% is of interest since it is known that such content of halogen doesn't worsen flexural properties of rubber.

Halide modification of polymers and natural rubber in particular together with obtaining of halogen-containing polymers with a help of synthesis is one of intensively developing direction in the field of obtaining chlorine-containing polymers. In result of carrying out halide modification of polymers, which have technologically smoothly, large capacity industrial production, elastomer materials and composites are managed to obtain with wide complex of a new specific properties: high adhesion, fire-, oil-, gasoline-, heat resistance, ozone resistance, incombustibility, resistance to influence of corrosive environments and microorganisms, high strength, gas permeability, etc.

10.2   RESULTS AND DISCUSSION

In earlier carried out works on haloid mechanochemical modification of synthetic analog of NR, isoprene SKI-3 [1–3] rubber, it was established that the size of the molecular weight (M_η) and the contents fraction gel (C_g, %) have essential impact on depth of course of reaction halogen accession to thermomechanical activated macromolecules of rubber.

The purpose of this work is studying of features of mechanochemical transformations of NR in the course of its processing in a two-rotor high-speed rubber-mixer for identification of the most optimum areas of carrying out haloid mechanochemical modification.

At the first stage we determined the size of M_η and the contents gel fraction of samples of NR of main suppliers of this natural polymer to the world market.

Certain structural parameters of NR and the producer country are specified in Table 10.1.

TABLE 10.1 Structural Parameters of NR

Type of rubber	[η]	$M_n \times 10^{-4}$	C_g, %
SVR 3L Vietnam	6.27	138	19.5
NKHC Cameroon	5.9	127.5	13.3
SMR GP Malaysia	6.4	144	2.5

From the data provided in the table it is visible that rubber of the Vietnamese production (SVR 3L) [4] most fully meets earlier designated requirements for M_η and the contents of gel fraction.

Further the process of conversion of natural rubber of the SVR 3L brand (Vietnam) was studied and structural parameters of samples of the processed natural rubber and property of elastomeric compositions on their basis are determined.

Mechanical processing of NR was carried out on a laboratory two-rotor rubber-mixer of type RVSD-01–60 (with friction 1:1,5). Duration of machining of SVR 3L rubber samples was 5, 10, 20, 30, 40, 50 and 60 min. Skilled samples of rubber were overworked in a self-heating mode. There were studied M_η change, the contents gel fraction and temperatures of rubbers depending on duration of mechanical influence.

Results of research of thermomechanochemical transformations of SVR 3L rubber are presented on Fig. 10.1.

From the provided data it is visible that at thermomechanical processing of NR mechanical degradation proceed generally at an initial stage of processing (up to 10 min) in the field of rather low temperatures (from 20 to 110 °C). In this time interval change of molecular weight from 140 to $68 \cdot 10^4$ and contents gel fraction from 20 to 5% is observed.

At processing times from 10 to 60 min little change of M_η from 68 to 50×10^4, and gel fraction from 5 to 3% is observed. Temperature of processed rubber in the range from 10 till 60 min of processing changes from 110 to 130 °C.

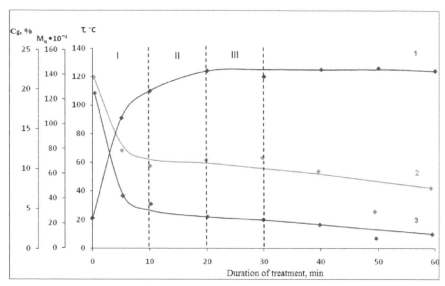

FIGURE 10.1   Temperature (1), average molecular weight (2) and the contents fraction gel (3) depending on duration of treatment of mechanical processing of SVR 3L (Vietnam).

By consideration of the results received at processing of NR from 10 till 60 min, it is possible to assume that thermo-mechanoactivation processes without a rupture of macromolecules and the partial (small) thermally activated destruction of macromolecules of rubber and its gel fraction generally prevail here.

For the purpose of studying of influence of structural parameters of NR samples processed in various temperature time intervals on properties of rubber compounds and rubbers on their basis, within the standard filled compounding for NR [5] rubber mixes were made and their vulcanizing properties and physicomechanical properties of vulcanizers were defined. Results of the conducted researches are given in Table 10.2.

From the data provided in Table 10.2 it is visible that with increase of processing time plasticity of rubber mixes increases caused by reduction the M out going and M min, connected with decrease of rubber molecular weight to course of degradant processes. Essential influence on such vulcanizing properties as time of the beginning of curing, optimum time of curing and speed of curing isn't observed. Some fall of value of the maximum torque (M_{max}) with increase in overtime of NR processing is noticeable that points to reduction of strength properties of vulcanizers of rubber samples.

TABLE 10.2 Influence of Duration of Treatment SVR 3L on Rubber Compounds Vulcanizing Properties and Physicomechanical Properties of Vulcanizers.

Index	Duration of treatment SVR 3L, min.							
	0	5	10	20	30	40	50	60
Rotational moment, dN*m								
M_{min}	6.7	6.5	6	6	3.5	2.8	2.5	2.5
M_{max}	29.7	28	27.5	27	24.5	22.2	22.5	21.5
Initial time of vulcanization, min	2	1.5	2	2.2	2.5	1.5	2.5	2.5
Optimum vulcanization time, min	7	7	7	8	8	7	8	8
Vulcanization rate, %/min	20	18.2	20	17.2	18.2	18.2	18.2	18.2
Conventional tensile strength 200%, MPa	1.6	1.5	1.1	1.3	1.2	1.1	1.1	1.1
Conventional tensile strength 500%, MPa	17.3	16.5	15.4	16.6	14.4	13.6	13.6	12.8
Conventional tensile strength, MPa	20.6	20.5	20.2	19.3	16.3	15	14.2	13.3
Conventional breaking elongation, %	675	660	620	610	600	600	575	570

By consideration of physicomechanical properties at increase in overtime of rubber falling of strength characteristics (conditional tension is revealed at 200%, 500% and conditional durability at stretching), insignificant decrease in relative lengthening and increase in relative residual lengthening after a gap are observed. It should be noted that till 30 min of processing inclusive, change of the parameters specified above is insignificant while at overtime, 40, 50 and 60 min their sharp change is observed.

On the basis of the critical data analysis obtained as a result of made by us experiment three areas of mechanochemical haloid modification of natural SVR 3L rubber are allocated:

- Area I–rubber overtime from 0 to 10 min (mechanodestruction of macromolecules and rubber gel fraction);
- Area II–rubber overtime from 10 to 20 min (mechanoactivation and thermally activated mechanodestruction);

- Area III-rubber overtime from 20 to 30 min (prevailing mechanoactivation and somewhat thermally activated mechanodestruction).

The most probable processes proceeding in the above-stated temperature time intervals, are presented on the following schemes:

I: mechanodestruction process

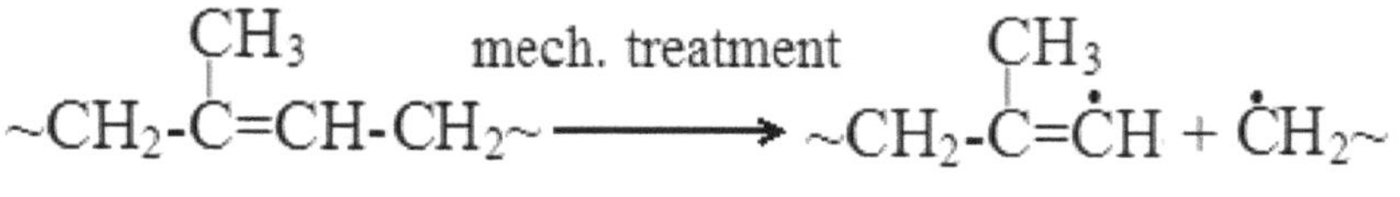

$$\sim CH_2\text{-}\underset{\underset{CH_3}{|}}{C}=CH\text{-}CH_2\sim \quad \xrightarrow{\text{mech. treatment}} \quad \sim CH_2\text{-}\underset{\underset{CH_3}{|}}{C}=\dot{C}H + \dot{C}H_2\sim$$

II: mechanodestruction and mechanoactivation (change of valent corners of the C-C communication without its gap)

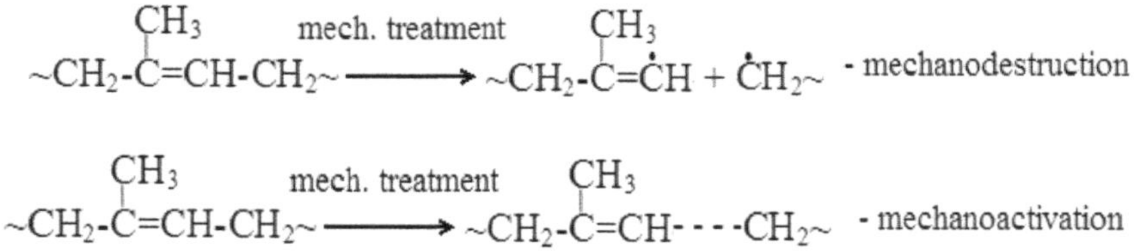

$$\sim CH_2\text{-}\underset{\underset{CH_3}{|}}{C}=CH\text{-}CH_2\sim \quad \xrightarrow{\text{mech. treatment}} \quad \sim CH_2\text{-}\underset{\underset{CH_3}{|}}{C}=\dot{C}H + \dot{C}H_2\sim \quad \text{- mechanodestruction}$$

$$\sim CH_2\text{-}\underset{\underset{CH_3}{|}}{C}=CH\text{-}CH_2\sim \quad \xrightarrow{\text{mech. treatment}} \quad \sim CH_2\text{-}\underset{\underset{CH_3}{|}}{C}=CH\text{-}\cdots\text{-}CH_2\sim \quad \text{- mechanoactivation}$$

III: mechanoactivation

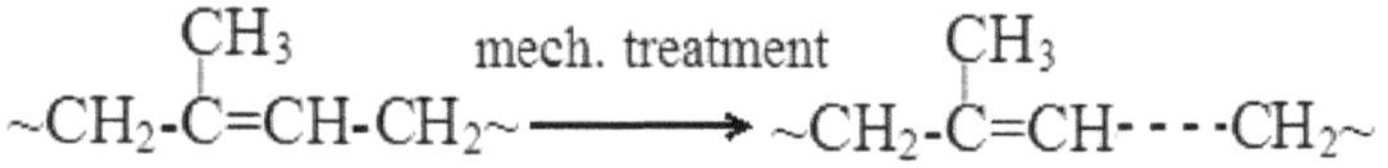

$$\sim CH_2\text{-}\underset{\underset{CH_3}{|}}{C}=CH\text{-}CH_2\sim \quad \xrightarrow{\text{mech. treatment}} \quad \sim CH_2\text{-}\underset{\underset{CH_3}{|}}{C}=CH\text{-}\cdots\text{-}CH_2\sim$$

10.3   CONCLUSIONS

Thus the conducted researches allowed to establish nature of change of structural parameters of natural SVR 3L rubber (M_η and the contents gel fraction) depending on duration of mechanical processing of polymer, to study the influence of these parameters on properties of elastomeric compositions on the basis of rubbers subjected thermo-mechanical influence. Taking into account the nature of the proceeding mechanochemical processes observed at processing of natural rubber, and also properties of elastomeric compositions on the basis of these rubbers the most accept-able temperature time intervals of carrying out mechanochemical haloid modification are defined. It is supposed that processing of rubbers in the above-stated temperature time intervals in the presence of the chlorine-containing modifier has to reveal influence of mechanical degradation and mechanoactivation processes on depth of reaction of halogenation.

KEYWORDS

- **Caoutchouc**
- **Elastomer**
- **Mechanical chemistry**
- **Rubber**
- **Rubber compound**
- **Technology**

REFERENCES

1. Andriasyan, Yu. O., Kornev, A. E. & Gyulbekyan, A. L. (2001). Theses of the Conference Report IX RAS. *Degradation and Stabilization of Polymers*. Moscow. 11.
2. Andriasyan, Yu. O., Popov, A. A., Gyulbekyan, A. L. & Kornev, A. E. (2002). *Caoutchouc and Rubber, 3*, 4–6.
3. Andriasyan, Yu. O., Popov, A. A., Gyulbekyan, A. L. & Kornev, A. E. (2002). *Caoutchouc and Rubber, 4*, 18–20.
4. Dumnov, S. E., Mikhailov, I. A., Andriasyan, Yu. O., Popov, A. A., Kashiricheva, I. I. & Kornev, A. E. Theses of Reports. Seventh Annual International Conference of Biochemical Physics RAS–Colleges (Moscow 2007). 102–104.
5. Manual of Rubberier. M.: Chemistry, 1971, 608.

CHAPTER 11

POLYMER/CARBON NANOTUBE MEMBRANE FILTRATION

A. K. HAGHI and G. E. ZAIKOV

CONTENTS

ABSTRACT

Membrane filtration is an important technology for ensuring the purity, safety and/or efficiency of the treatment of water or effluents. In this study, various types of membranes are reviewed, first. After that, the states of the computational methods are applied to membranes processes. Many studies have focused on the best ways of using a particular membrane process. But, the design of new membrane systems requires a considerable amount of process development as well as robust methods. Monte Carlo and molecular dynamics methods can specially provide a lot of interesting information for the development of polymer/carbon nanotube membrane processes.

11.1 MEMBRANES FILTRATION

Membrane filtration is a mechanical filtration technique, which uses an absolute barrier to the passage of particulate material as any technology currently available in water treatment. The term "membrane" covers a wide range of processes, including those used for gas/gas, gas/liquid, liquid/liquid, gas/solid, and liquid/solid separations. Membrane production is a large-scale operation. There are two basic types of filters: depth filters and membrane filters.

Depth filters have a significant physical depth and the particles to be maintained are captured throughout the depth of the filter. Depth filters often have a flexuous three-dimensional structure, with multiple channels and heavy branching so that there is a large pathway through which the liquid must flow and by which the filter can retain particles. Depth filters have the advantages of low cost, high through put, large particle retention capacity, and the ability to retain a variety of particle sizes. However, they can endure from entrainment of the filter medium, uncertainty regarding effective pore size, some ambiguity regarding the overall integrity of the filter, and the risk of particles being mobilized when the pressure differential across the filter is large.

The second type of filter is the membrane filter, in which depth is not considered momentous. The membrane filter uses a relatively thin material with a well-defined maximum pore size and the particle retaining effect takes place almost entirely at the surface. Membranes offer the advantage of having well-defined effective pore sizes, can be integrity tested more easily than depth filters, and can achieve more filtration of much smaller particles. They tend to be more expensive than depth filters and usually cannot achieve the throughput of a depth filter. Filtration technology has developed a well-defined terminology that has been well addressed by commercial suppliers.

The term membrane has been defined in a number of ways. The most appealing definitions to us are the following:

"A selective separation barrier for one or several components in solution or suspension"[19]."A thin layer of material that is capable of separating materials as a function of their physical and chemical properties when a driving force is applied across the membrane."

Membranes are important materials which form part of our daily lives. Their long history and use in biological systems has been extensively studied throughout the scientific field. Membranes have proven themselves as promising separation candidates due to advantages offered by their high stability, efficiency, low energy requirement and ease of operation. Membranes with good thermal and mechanical stability combined with good solvent resistance are important for industrial processes [1].

The concept of membrane processes is relatively simple but nevertheless often unknown. Membranes might be described as conventional filters but with much finer mesh or much smaller pores to enable the separation of tiny particles, even molecules. In general, one can divide membranes into two groups: porous and nonporous. The former group is similar to classical filtration with pressure as the driving force; the separation of a mixture is achieved by the rejection of at least one component by the membrane and passing of the other components through the membrane (see Fig. 11.1). However, it is important to note that nonporous membranes do not operate on a size exclusion mechanism.

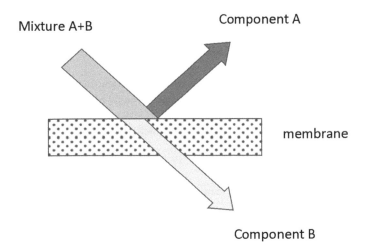

FIGURE 11.1   Basic principle of porous membrane processes.

Membrane separation processes can be used for a wide range of applications and can often offer significant advantages over conventional separation such as distillation and adsorption since the separation is based on a physical mechanism. Compared to conventional processes, therefore, no chemical, biological, or thermal change of the component is involved for most membrane processes. Hence membrane separation is particularly attractive to the processing of food, beverage, and bioproducts where the processed products can be sensitive to temperature (vs. distillation) and solvents (vs. extraction).

Synthetic membranes show a large variety in their structural forms. The material used in their production determines their function and their driving forces. Typically the driving force is pressure across the membrane barrier (see Table 11.1) [2–4]. Formation of a pressure gradient across the membrane allows separation in a bolter-like manner. Some other forms of separation that exist include charge effects and solution diffusion. In this separation, the smaller particles are allowed to pass through as per Meates whereas the larger molecules (macromolecules) are retained. There tension or permeation of these species is ordained by the pore architecture as well as poresizes of the membrane employed. Therefore based on the poresizes, these pressure driven membranes can be divided into reverse osmosis (RO), nanofiltration (NF), ultrafiltration (UF), and microfiltration (MF), are already applied on an industrial scale to food and bio product processing [5–7].

TABLE 11.1   Driving Forces and their Membrane Processes

Driving force	Membrane process
Pressure difference	Microfiltration, Ultrafiltration, Nanofiltration, Reverse osmosis
Chemical potential difference	Pervaporation, Perstraction, Dialysis, Gas separation, Vapor permeation, Liquid Membranes
Electrical potential difference	Electrodialysis, Membrane electrophoresis, Membrane electrolysis
Temperature difference	Membrane distillation

11.1.1 MICROFILTRATION (MF) MEMBRANES

MF membranes have the largest pore sizes and thus useless pressure. They involve removing chemical and biological species with diameters ranging between 100 to 10,000 nm and components smaller than this, pass through as permeates. MF is primarily used to separate particles and bacteria from other smaller solutes [4].

11.1.2 ULTRAFILTRATION (UF) MEMBRANES

UF membranes operate within the parameters of the micro and nano filtration membranes. Therefore UF membranes have smaller pores as compared to MF membranes. They involve retaining macromolecules and colloids from solution, which range between 2–100 nm and operating pressures between 1 and 10 bar, for example, large organic molecules and proteins. UF is used to separate colloids such as proteins from small molecules such as sugars and salts [4].

11.1.3 NANOFILTRATION (NF) MEMBRANES

NF membranes are distinguished by their poresizes of between 0.5-2 nm and operating pressures between 5 and 40 bar. They are mainly used for the removal of small organic molecules and di- and multivalentions. Additionally, NF membranes have surface charges that make them suitable for retaining ionic pollutants from solution. NF is used to achieve separation between sugars, other organic molecules, and multivalent salts on the one hand from monovalent salts and water on the other. Nanofiltration, however, does not remove dissolved compounds [4].

11.1.4 REVERSE OSMOSIS (RO) MEMBRANES

RO membranes are dense semipermeable membranes mainly used for desalination of sea water [38]. Contrary to MF and UF membranes, RO membranes have no distinct pores. As a result, high pressures are applied to increase the permeability of the membranes [4]. The properties of the various types of membranes are summarized in Table 11.2

Table 11.2. Summary of Properties of Pressure Driven Membranes [4]

	MF	UF	NF	RO
Permeability(L/h. m².bar)	1000	10–1000	1.5–30	0.05–1.5
Pressure (bar)	0.1–2	0.1–5	3–20	5–1120
Poresize (nm)	100–10000	2–100	0.5–2	0.5
Separation Mechanism	Sieving	Sieving	Sieving, charge effects	Solution diffusion
Applications	Removal of bacteria	Removal of bacteria, fungi, viruses	Removal of multivalentions	desalination

The NF membrane is a type of pressure-driven membrane with properties in between RO and UF membranes. NF offers several advantages such as low operation pressure, high flux, high retention of multivalent anion salts and an organic molecular above 300, relatively low investment and low operation and maintenance costs. Because of these advantages, the applications of NF worldwide have increased [8]. In recent times, research in the application of nanofiltration techniques has been extended from separation of aqueous solutions to separation of organic solvents to homogeneous catalysis, separation of ionic liquids, food processing, etc. [9].

Figure 11.2 presents a classification on the applicability of different membrane separation processes based on particle or molecular sizes. RO process is often used for desalination and pure water production, but it is the UF and MF that are widely used in food and bioprocessing.

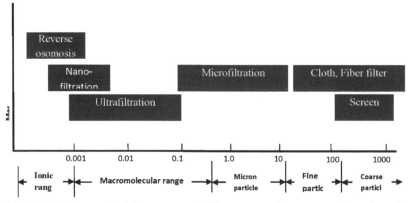

FIGURE 11.2   The applicability ranges of different separation processes based on sizes.

While MF membranes target on the microorganism removal, and hence are given the absolute rating, namely, the diameter of the largest pore on the membrane surface, UF/NF membranes are characterized by the nominal rating due to their early applications of purifying biological solutions. The nominal rating is defined as the molecular weight cut-off (MWCO) that is the smallest molecular weight of species, of which the membrane has more than 90% rejection (see later for definitions). The separation mechanism in MF/UF/NF is mainly the size exclusion, which is indicated in the nominal ratings of the membranes. The other separation mechanism includes the electrostatic interactions between solutes and membranes, which depends on the surface and physiochemical properties of solutes and membranes [5].Also, The principal types of membrane are shown schematically in Fig. 11.4 and are described briefly in Section 12.2.

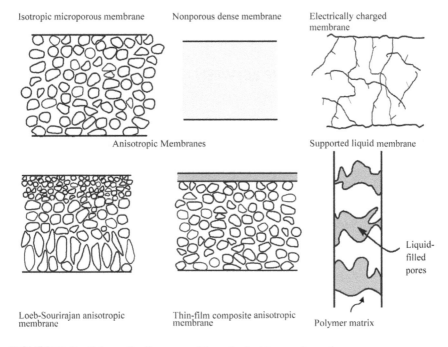

FIGURE 11.3 Schematic diagrams of the principal types of membranes.

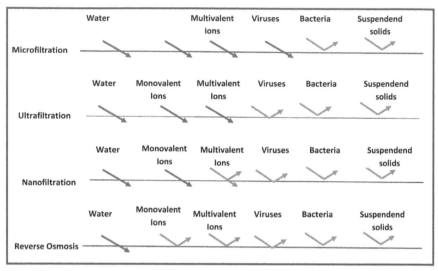

FIGURE 11.4   Membrane process characteristics.

11.2   THE RELATIONSHIP BETWEEN NANOTECHNOLOGY AND FILTRATION

Nowadays, nano materials have become the most interested topic of materials research and development due to their unique structural properties (unique chemical, biological, and physical properties as compared to larger particles of the same material) that cover their efficient uses in various fields, such as ion exchange and separation, catalysis, bio molecular isolation and purification as well as in chemical sensing [10]. However, the understanding of the potential risks (health and environmental effects) posed by nano materials hasn't increased as rapidly as research has regarding possible applications.

One of the ways to enhance their functional properties is to increase their specific surface area by the creation of a large number of nano-structured elements or by the synthesis of a highly porous material.

Classically, porous matter is seen as material containing three-dimensional voids, representing translational repetition, while no regularity is necessary for a material to be termed "porous." In general, the pores can be classified into two types: open pores, which connect to the surface of the material, and closed pores which are isolated from the outside. If the material exhibits mainly open pores, which can be easily transpired, then one can consider its use in functional applications such as adsorption, catalysis and sensing. In turn, the closed pores can be used in sonic and thermal insulation, or lightweight structural applications. The use of porous materials offers also new opportunities in such areas as coverage chemistry, guest–host synthesis and molecular manipulations and reactions for manufacture of nano particles, nano wires and other quantum nanostructures. The International Union of Pure and Applied Chemistry (IUPAC) defines porosity scales as follows:

- Microporous materials 0–2-nm pores
- Mesoporous materials 2–50-nm pores
- Macroporous materials >50-nm pores

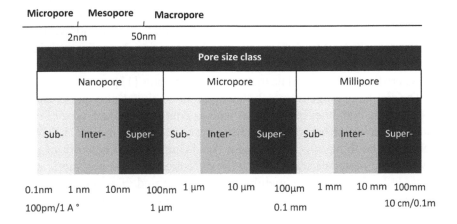

FIGURE 11.5 New pore size classification as compared with the current IUPAC nomenclature.

This definition, it should be noted, is somewhat in conflict with the definition of nano scale objects, which typically have large relative porosities (>0.4), and pore diameters between 1 and 100 nm. In order to classify porous materials according to the size of their pores the sorption analysis is one of the tools often used. This tool is based on the fact that pores of different sizes lead to totally different characteristics in sorption isotherms. The correlation between the vapor pressure and the pore size can be written as the Kelvin equation:

$$r_p\left(\frac{p}{p_0}\right) = \frac{2\gamma V_L}{RT\ln\left(\frac{p}{p_0}\right)} + t\left(\frac{p}{p_0}\right) \qquad (1)$$

Therefore, the isotherms of microporous materials show a steep increase at very low pressures (relative pressures near zero) and reach a plateau quickly. Mesoporous materials are characterized by a so-called capillary doping step and a hysteresis (a discrepancy between adsorption and desorption). Macroporous materials show a single or multiple adsorption steps near the pressure of the standard bulk condensed state (relative pressure approaches one) [10].

Nanoporous materials exuberate in nature, both in biological systems and in natural minerals. Some nanoporous materials have been used industrially for a longtime. Recent progress in characterization and manipulation on the nanoscale has led to noticeable progression in understanding and making a variety of nanoporous materials: from the merely opportunistic to directed design. This is most strikingly the case in the creation of a wide variety of membranes where control over pore size is increasing dramatically, often to atomic levels of perfection, as is the ability to modify physical and chemical characteristics of the materials that make up the pores [11].

The available range of membrane materials includes polymeric, carbon, silica, zeolite and other ceramics, as well as composites. Each type of membrane can have a different porous structure, as illustrated in Fig. 11.6. Membranes can be thought of as having a fixed (immovable) network of pores in which the molecule travels, with the exception of most polymeric membranes [12, 13]. Polymeric membranes are composed of an amorphous mix of polymer chains whose interactions involve mostly Van der Waals forces. However, some polymers manifest a behavior that is consistent with the idea of existence of opened pores within their matrix. This is especially true for high free volume, high permeability polymers, as has been proved by computer modeling, low activation energy of diffusion, negative activation energy of permeation, solubility controlled permeation [14, 15]. Although polymeric membranes have often been viewed as nonporous, in the modeling framework discussed here it is convenient to consider them nonetheless as porous. Glassy polymers have pores that can be considered as 'frozen' over short times scales, while rubbery polymers have dynamic fluctuating pores (or more correctly free volume elements) that move, shrink, expand and disappear [16].

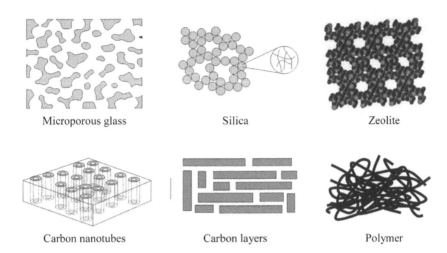

| Microporous glass | Silica | Zeolite |

| Carbon nanotubes | Carbon layers | Polymer |

FIGURE 11.6   Porous structure within various types of membranes

Three nanotechnologies that are often used in the filtering processes and show great potential for applications in remediation are:

1. Nanofiltration (and its sibling technologies: reverse osmosis, ultra filtration, and microfiltration), is a fully developed, commercially available membrane technology with a large number of vendors. Nanofiltration relies on the ability of membranes to discriminate between the physical size of particles or species in a mixture or solution and is primarily used for water pretreatment, treatment, and purification). There are almost 600 companies in worldwide which offering membrane systems.

2. Electro spinning is a process used by the nanofiltration process, in which fibers are stretched and elongated down to a diameter of about 10 nm. The modified nanofibers that are produced are particularly useful in the filtration process as an ultra-concentrated filter with a very large surface area. Studies have found that electro spun nanofibers can capture metallic ions and are continually effective through refiltration.

3. Surface modified membrane is a term used for membranes with altered makeup and configuration, though the basic properties of their underlying materials remain intact.

11.3   TYPES OF MEMBRANES

As it mentioned, membranes have achieved a momentous place in chemical technology and are used in a broad range of applications. The key property that is exploited is the ability of a membrane to control the permeation rate of a chemical species through the membrane. In essence, a membrane is nothing more than a discrete, thin interface that moderates the permeation of chemical species in contact with it. This interface may be molecularly homogeneous, that is completely uniform in composition and structure or it may be chemically or physically heterogeneous for example, containing holes or pores of finite dimensions or consisting of some form of layered structure. A normal filter meets this definition of a membrane, but, generally, the term filter is usually limited to structures that separate particulate suspensions larger than 1–10 µm [17].

The preparation of synthetic membranes is however a more recent invention which has received a great audience due to its applications [18]. Membrane technology like most other methods has undergone a developmental stage, which has validated the technique as a cost-effective treatment option for water. The level of performance of the membrane technologies is still developing and it is stimulated by the use of additives to improve the mechanical and thermal properties, as well as the permeability, selectivity, rejection and fouling of the membranes [19]. Membranes can be fabricated to possess different morphologies. However, most membranes that have found practical use are mainly of asymmetric structure. Separation in membrane processes takes place as a result of difference in the transport rates of different species through the membrane structure, which is usually polymer i-core ceramic [20].

The versatility of membrane filtration has allowed their use in many processes where their properties are suitable in the feeds stream. Although membranes separation does not provide the three ultimate solution to water treatment it can be economically connect to conventional treatment technologies by modifying and improving certain properties [21]

The performance of any polymeric membrane in a given process is highly dependent on both the chemical structure of the matrix and the physical arrangement of the membrane [22] moreover, the structural impeccability of a membrane is very important once it determines its permeation and de selectivity efficiency. As such, polymembranes should be seen as much more than just serving filter, but as intrinsic complex structures which can either be homogenous (isotropic) or heterogeneous (anisotropic), porous or dense, liquid or solid, orange or inorganic [22, 23]

11.3.1 ISOTROPIC MEMBRANES

Isotropic membranes are typically homogeneous/uniform in composition and structure. They are divided into three subgroups, namely: microporous, dense and electrically charged membranes [20]. Isotropic microporous membranes have evenly distributed pores (Fig. 11.7a) [27]. Their pore diameters range between 0.01–10 µm and operate by the sieving mechanism. The microporous membranes are mainly prepared by the phase inversion method albeit other methods can be used. Conversely, isotropic dense membranes do not have pores and as a result they tend to be thicker than the microporous membranes (Fig. 11.7b). Solutes are carried through the membrane by diffusion under a pressure, concentration or electrical potential gradient. Electrically charged membranes can either be porous or nonporous. However in most cases they are finely microporous with pore walls containing charged ions (Fig. 11.7c) [20, 28].

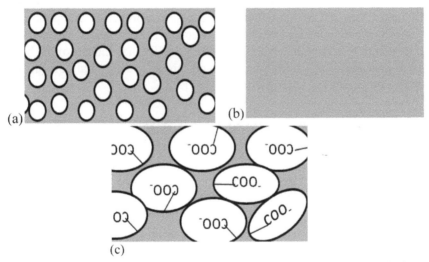

FIGURE 11.7 Schematic diagrams of isotropic membranes: (a) microporous; (b) dense; and (c) electrically charged membranes.

11.3.2 ANISOTROPIC MEMBRANES

Anisotropic membranes are often referred to as Loeb-Sourirajan, based on the scientists who first synthesized them [24, 25]. They are the most widely used membranes in industries. The transport rate of a species through a membrane is inversely proportional to the membrane thickness. The membrane should be as thin as possible due to high transport rates are eligible in membrane separation processes for economic reasons. Contractual film fabrication technology limits manufacture of mechanically strong, defect-free films to thicknesses of about 20 μm. The development of novel membrane fabrication techniques to produce anisotropic membrane structures is one of the major breakthroughs of membrane technology. Anisotropic membranes consist of an extremely thin surface layer supported on a much thicker, porous substructure. The surface layer and its substructure may be formed in a single operation or separately [17]. They are represented by nonuniform structures, which consist of a thin active skin layer and a highly porous support layer. The active layer enjoins the efficiency of the membrane, where as the porous up port layer influences the mechanical stability of the membrane. An isotropic membranes can be classified into two groups, namely: (i) integrally skinned membranes where the active layer is formed from the same substance as the supporting layer, (ii) composite membranes where the polymer of the active layer differs from that of the supporting sublayer [25]. In composite membranes, the layers are usually made from different polymers. The separation properties and permeation rates of the membrane are determined particularly by the surface layer and the substructure functions as a mechanical support. The advantages of the higher fluxes provided by anisotropic membranes are so great that almost all commercial processes use such membranes [17].

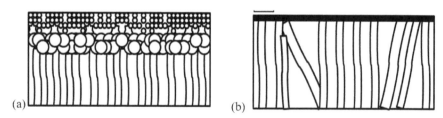

FIGURE 11.8   Schematic diagrams of anisotropic membranes: (a) Loeb-Sourirajan, and (b) thin film composite membranes.

11.3.3   POROUS MEMBRANE

In Knudsen diffusion (Fig. 11.9a), the pore size forces the penetrant molecules to collide more frequently with the pore wall than with other incisive species [26]. Except for some special applications as membrane reactors, Knudsen-selective membranes are not commercially attractive because of their low selectivity [27]. In surface diffusion mechanism (Fig. 11.9 b), the pervasive molecules adsorb on the surface of the pores so move from one site to another of lower concentration. Capillary condensation (Fig. 11.9c) impresses the rate of diffusion across the membrane. It occurs when the pore size and the interactions of the penetrant with the pore walls induce penetrant condensation in the pore [28]. Molecular-sieve membranes in Fig. 11.9d have gotten more attention because of their higher productivities and selectivity than solution-diffusion membranes. Molecular sieving membranes are means to polymeric membranes. They have ultra microporous (<7Å) with sufficiently small pores to barricade some molecules, while allowing others to pass through. Although they have several advantages such as permeation performance, chemical and thermal stability, they are still difficult to process because of some properties like fragile. Also they are expensive to fabricate.

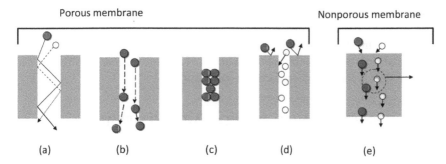

FIGURE 11.9 Schematic representation of membrane-based gas separations. (a) Knudsen-flow separation, (b) surface-diffusion, (c) capillary condensation, (d) molecular-sieving separation, and (e) solution-diffusion mechanism.

11.3.4 NONPOROUS (DENSE) MEMBRANE

Nonporous, dense membranes consist of a dense film through which permeants are transported by diffusion under the driving force of a pressure, concentration, or electrical potential gradient. The separation of various components of a mixture is related directly to their relative transport rate within the membrane, which is determined by their diffusivity and solubility in the membrane material. Thus, nonporous, dense membranes can separate permeants of similar size if the permeant concentrations in the membrane material differ substantially. Reverse osmosis membranes use dense membranes to perform the separation. Usually these membranes have an anisotropic structure to improve the flux[17].

The mechanism of separation by nonporous membranes is different from that byporous membranes. The transport through nonporous polymeric membranes is usually described by a solution–diffusion mechanism (Fig. 11.9e). The most current commercial polymeric membranes operate according to the solution–diffusion mechanism. The solution–diffusion mechanism has three steps: (i) the absorption or adsorption at the upstream boundary, (ii) activated diffusion through the membrane, and (iii) desorption or evaporation on the other side. This solution–diffusion mechanism is driven by a difference in the thermodynamic activities existing at the upstream and downstream faces of the membrane as well as the intermolecular forces acting between the permeating molecules and those making up the membrane material.

The concentration gradient causes the diffusion in the direction of decreasing activity. Differences in the permeability in dense membranes are caused not only by diffusivity differences of the various species but also by differences in the physicochemical interactions of the species within the polymer. The solution–diffusion model assumes that the pressure within a membrane is uniform and that the chemical potential gradient across the membrane is expressed only as a concentration gradient. This mechanism controls permeation in polymeric membranes for separations.

11.4   CARBON NANOTUBES-POLYMER MEMBRANE

Iijima discovered carbon nanotubes (CNTs) in 1991 and it was really a revolution in nanoscience because of their distinguished properties. CNTs have the unique electrical properties and extremely high thermal conductivity [29, 30] and high elastic modulus (>1 TPa), large elastic strain-up to 5%, and large breaking strain-up to 20%. Their excellent mechanical properties could lead to many applications[31]. For example, with their amazing strength and stiffness, plus the advantage of lightness, perspective future applications of CNTs are in aerospace engineering and virtual biodevices [32].

CNTs have been studied worldwide by scientists and engineers since their discovery, but a robust, theoretically precise and efficient prediction of the mechanical properties of CNTs has not yet been found. The problem is, when the size of an object is small to nanoscale, their many physical properties cannot be modeled and analyzed by using constitutive laws from traditional continuum theories, since the complex atomistic processes affect the results of their macroscopic behavior. Atomistic simulations can give more precise modeled results of the underlying physical properties. Due to atomistic simulations of a whole CNT are computationally infeasible at present, a new atomistic and continuum mixing modeling method is needed to solve the problem, which requires crossing the length and time scales. The research here is to develop a proper technique of spanning multiscales from atomic to macroscopic space, in which the constitutive laws are derived from empirical atomistic potentials which deal with individual interactions between single atoms at the microlevel, whereas Cosserat continuum theories are adopted for a shell model through the application of the Cauchy-Born rule to give the properties which represent the averaged behavior of large volumes of atoms at the macrolevel [33, 34]. Since experiments of CNTs are relatively expensive at present, and often unexpected manual errors could be involved, it will be very helpful to have a mature theoretical method for the study of mechanical properties of CNTs. Thus, if this research is successful, it could also be a reference for the research of all sorts of research at the nanoscale, and the results can be of interest to aerospace, biomedical engineering [35].

Subsequent investigations have shown that CNTs integrate amazing rigid and tough properties, such as exceptionally high elastic properties, large elastic strain, and fracture strain sustaining capability, which seem inconsistent and impossible in the previous materials. CNTs are the strongest fibers known. The Young's Modulus of SWNT is around 1 TPa, which is 5 times greater than steel (200 GPa) while the density is only $1.2 \sim 1.4$ g/cm^3. This means that materials made of nanotubes are lighter and more durable.

Beside their well-known extra high mechanical properties, single-walled carbon nanotubes (SWNTs) offer either metallic or semiconductor characteristics based on the chiral structure of fullerene. They possess superior thermal and electrical properties so SWNTs are regarded as the most promising reinforcement material for the next generation of high performance structural and multifunctional composites, and evoke great interest in polymer based composites research. The SWNTs/polymer composites are theoretically predicted to have both exceptional mechanical and functional properties, which carbon fibers cannot offer [36].

11.4.1 CARBON NANOTUBES

Nanotubular materials are important "building blocks" of nanotechnology, in particular, the synthesis and applications of CNTs [37, 39]. One application area has been the use of carbon nanotubes for molecular separations, owing to some of their unique properties. One such important property, extremely fast mass transport of molecules within carbon nanotubes associated with their low friction inner nanotube surfaces, has been demonstrated via computational and experimental studies [40, 41]. Furthermore, the behavior of adsorbate molecules in nano-confinement is fundamentally different than in the bulk phase, which could lead to the design of new sorbents [42].

Finally, their one-dimensional geometry could allow for alignment in desirable orientations for given separation devices to optimize the mass transport. Despite possessing such attractive properties, several intrinsic limitations of carbon nanotubes inhibit their application in large scale separation processes: the high cost of CNT synthesis and membrane formation (by micro fabrication processes), as well as their lack of surface functionality, which significantly limits their molecular selectivity [43]. Although outer-surface modification of carbon nanotubes has been developed for nearly two decades, interior modification via covalent chemistry is still challenging due to the low reactivity of the inner-surface. Specifically, forming covalent bonds at inner walls of carbon nanotubes requires a transformation from sp^2 to sp^3 hybridization. The formation of sp^3 carbon is energetically unfavorable for concave surfaces [44].

Membrane is a potentially effective way to apply nanotubular materials in industrial-scale molecular transport and separation processes. Polymeric membranes are already prominent for separations applications due to their low fabrication and operation costs. However, the main challenge for using polymer membranes for future high-performance separations is to overcome the tradeoff between permeability and selectivity. A combination of the potentially high throughput and selectivity of nanotube materials with the process ability and mechanical strength of polymers may allow for the fabrication of scalable, high-performance membranes [45, 46].

11.4.2 STRUCTURE OF CARBON NANOTUBES

Two types of nanotubes exist in nature: multiwalled carbon nanotube) MWNTs(, which were discovered by Iijima in 1991 [39] and SWNTs, which were discovered by Bethune et al. in 1993 [47–48].

Single-wall nanotube has only one single layer with diameters in the range of 0.6–1 nm and densities of 1.33–1.40 g/cm^3 [49] MWNTs are simply composed of concentric SWNTs with an inner diameter is from 1.5 to 15 nm and the outer diameter is from 2.5 nm to 30 nm [50]. SWNTs have better defined shapes of cylinder than MWNT, thus MWNTs have more possibilities of structure defects and their nanostructure is less stable. Their specific mechanical and electronic properties make them useful for future high strength/modulus materials and nano devices. They exhibit low density, large elastic limit without breaking (of up to 20–30%strain before failure), exceptional elastic stiffness, greater than 1000 GPa and their extreme strength which is more than 20 times higher than a high-strength steel alloy. Besides, they also posses superior thermal and elastic properties: thermal stability up to 2800 °C in vacuum and up to 750 °C in air, thermal conductivity about twice as high as diamond, electric current carrying capacity 1000timeshigherthancopperwire [51]. The properties of CNTs strongly depend on the size and the chirality and dramatically change when SWCNTs or MWCNTs are considered [52].

CNTs are formed from pure carbon bonds. Pure carbons only have two covalent bonds: sp^2 and sp^3. The former constitutes graphite and the latter constitutes diamond. The sp^2 hybridization, composed of one s orbital and two p orbitals, is a strong bond within a plane but weak between planes. When more bonds come together, they form six-fold structures, like honeycomb pattern, which is a plane structure, the same structure as graphite [53].

Graphite is stacked layer by layer so it is only stable for one single sheet. Wrapping these layers into cylinders and joining the edges, a tube of graphite is formed, called nanotube [54].

Atomic structure of nanotubes can be described in terms of tube chirality, or helicity, which is defined by the chiral vector, and the chiral angle, θ. Figure 11.10 shows visualized cutting a graphite sheet along the dotted lines and rolling the tube so that the tip of the chiral vector touches its tail. The chiral vector, often known as the roll-up vector, can be described by the following equation [55]:

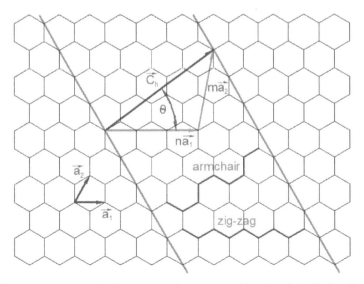

FIGURE 11.10 Schematic diagram showing how graphite sheet is 'rolled' to form CNT

$$C_h = na_1 + ma_2 \qquad (2)$$

As shown in Fig. 11.10, the integers (n, m) are the number of steps along the carbon bonds of the hexagonal lattice. Chiral angle determines the amount of "twist" in the tube. Two limiting cases exist where the chiral angle is at 0° and 30°. These limiting cases are referred to as ziz-zag (0°) and armchair (30°), based on the geometry of the carbon bonds around the circumference of the nanotube. The difference in armchair and zig-zag nanotube structures is shown in Fig. 11.11 In terms of the roll-up vector, the ziz-zag nanotube is (n, 0) and the armchair nanotube is (n, n). The roll-up vector of the nanotube also defines the nanotube diameter since the interatomic spacing of the carbon atoms is known.[36]

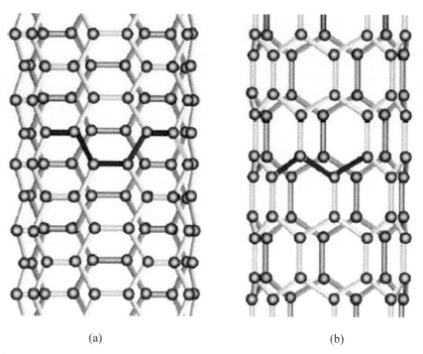

(a)   (b)

FIGURE 11.11   Illustrations of the atomic structure (a) an armchair and (b) a ziz-zag nanotube

Chiral vector C_h is a vector that maps an atom of one end of the tube to the other. C_h can be an integer multiple a_1 of a_2, which are two basis vectors of the graphite cell. Then we have $C_h = a_1 + a_2$, with integer n and m, and the constructed CNT is called a (n, m) CNT, as shown in Fig. 11.12. It can be proved that for armchair CNTs n=m, and for zigzag CNTs m=0.In Fig. 11.12, the structure is designed to be a (4,0) zigzag SWCNT.

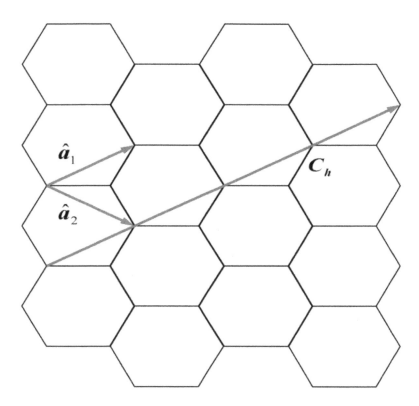

FIGURE 11.12   Basis vectors and chiral vector

MWCNT can be considered as the structure of a bundle of concentric SWCNTs with different diameters. The length and diameter of MWCNTs are different from those of SWCNTs, which means, their properties differ signifi-cantly. MWCNTs can be modeled as a collection of SWCNTs, provided the interlayer interactions are modeled by Van der Waals forces in the simulation. A SWCNT can be modeled as a hollow cylinder by rolling a graphite sheet as presented in Fig. 11.13.

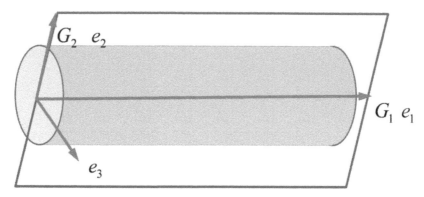

FIGURE 11.13   Illustration of a graphite sheet rolling to SWCNT

If a planar graphite sheet is considered to be an undeformed configuration, and the SWCNT is defined as the current configuration, then the relationship between the SWCNT and the graphite sheet can be shown to be:

$$e_1 = G_1, e_2 = R\sin\frac{G_2}{R}, e_3 = R\cos\frac{G_2}{R} - R \tag{3}$$

The relationship between the integer's n, m and the radius of SWCNT is given by:

$$R = a\sqrt{m^2 + mn + n^2} / 2\pi \tag{4}$$

where, and a_0 is the length of a nonstretched C-C bond which is 0.142 nm [56].

As a graphite sheet can be 'rolled' into a SWCNT, we can 'unroll' the SWCNT to a plane graphite sheet. Since a SWCNT can be considered as a rectangular strip of hexagonal graphite monolayer rolling up to a cylindrical tube, the general idea is that it can be modeled as a cylindrical shell, a cylinder surface, or it can pull-back to be modeled as a plane sheet deforming into curved surface in three-dimensional space. A MWCNT can be modeled as a combination of a series of concentric SWCNTs with interlayer inter-atomic reactions. Provided the continuum shell theory captures the deformation at the macrolevel, the inner microstructure can be described by finding the appropriate form of the potential function which is related to the position of the atoms at the atomistic level. Therefore, the SWCNT can be considered as a generalized continuum with microstructure[35].

11.4.3   CNT COMPOSITES

CNT composite materials cause significant development in nanoscience and nanotechnology. Their remarkable properties offer the potential for fabricating composites with substantially enhanced physical properties including conductivity, strength, elasticity, and toughness. Effective utilization of CNT in composite applications is dependent on the homogeneous distribution of CNTs throughout the matrix. Polymer-based nanocomposites are being developed for electronics applications such as thin-film capacitors in integrated circuits and solid polymer electrolytes for batteries. Research is being conducted throughout the world targeting the application of carbon nanotubes as materials for use in transistors, fuel cells, big TV screens, ultra-sensitive sensors, high-resolution Atomic Force Microscopy (AFM) probes, super-capacitor, transparent conducting film, drug carrier, catalysts, and composite material. Nowadays, there are more reports on the fluid transport through porous CNTs/polymer membrane.

11.4.4   STRUCTURAL DEVELOPMENT IN POLYMER/CNT FIBERS

The inherent properties of CNT assume that the structure is well preserved (large-aspect-ratio and without defects). The first step toward effective reinforcement of polymers using nano-fillers is to achieve a uniform dispersion of the fillers within the hosting matrix, and this is also related to the as-synthesized nano-carbon structure. Secondly, effective interfacial interaction and stress transfer between CNT and polymer is essential for improved mechanical properties of the fiber composite. Finally, similar to polymer molecules, the excellent intrinsic mechanical properties of CNT can be fully exploited only if an ideal uniaxial orientation is achieved. Therefore, during the fabrication of polymer/CNT fibers, four key areas need to be addressed and understood in order to successfully control the microstructural development in these composites. These are: (i) CNT pristine structure, (ii) CNT dispersion, (iii) polymer–CNT interfacial interaction, and (iv) orientation of the filler and matrix molecules (Fig. 11.14). Figure 11.14 Four major factors affecting the microstructural development in polymer/CNT composite fiber during processing [57].

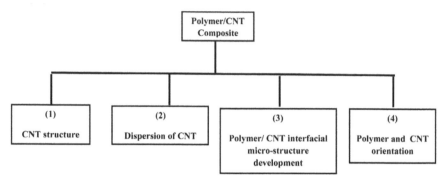

FIGURE 11.14   Four major factors affecting the microstructural development in polymer/ CNT composite fiber during processing.

Achieving homogenous dispersion of CNTs in the polymer matrix through strong interfacial interactions is crucial to the successful development of CNT/ polymer nanocomposite [58]. As a result, various chemical or physical modifications can be applied to CNTs to improve its dispersion and compatibility with polymer matrix. Among these approaches acid treatment is considered most convenient, in which hydroxyl and carboxyl groups generated would concentrate on the ends of the CNT and at defect sites, making them more reactive and thus better dispersed [59, 60].

The incorporation of functionalized CNTs into composite membranes are mostly carried out on flat sheet membranes[61, 62]. For considering the potential influences of CNTs on the physicochemical properties of dope solution [63] and change of membrane formation route originated from various additives [64], it is necessary to study the effects of CNTs on the morphology and performance.

11.4.5   GENERAL FABRICATION PROCEDURES FOR POLYMER/CNT FIBERS

In general, when discussing polymer/CNT composites, two major classes come to mind. First, the CNT nano-fillers are dispersed within a polymer at a specified concentration, and the entire mixture is fabricated into a composite. Secondly, as grown CNT are processed into fibers or films, and this macroscopic CNT material is then embedded into a polymer matrix [65]. The four major fiber-spinning methods (Fig. 11.15) used for polymer/CNT composites from both the solution and melt include dry-spinning [66], wet-spinning [67], dry-jet wet spinning (gel-spinning), and electro spinning [68]. An ancient solid-state spinning approach has been used for fabricating 100% CNT fibers from both forests and aerogels. Irrespective of the processing technique, in order to develop high-quality fibers many parameters need to be well controlled.

All spinning procedures generally involve:

(i) Fiber formation (ii)coagulation/gelation/solidification and (iii) drawing/alignment.

For all of these processes, the even dispersion of the CNT within the polymer solution or melt is very important. However, in terms of achieving excellent axial mechanical properties, alignment and orientation of the polymer chains and the CNT in the composite is necessary. Fiber alignment is accomplished in postprocessing such as drawing/annealing and is key to increasing crystallinity, tensile strength, and stiffness [69].

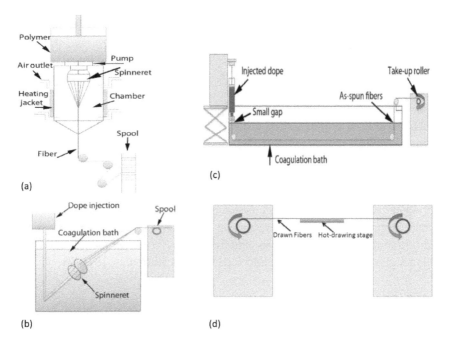

FIGURE 11.15 Schematics representing the various fiber processing methods (a) dry-spinning; (b) wet-spinning; (c) dry-jet wet or gel-spinning; and (d) postprocessing by hot-stage drawing.

11.5   COMPUTATIONAL METHODS

Computational approaches to obtain solubility and diffusion coefficients of small molecules in polymers have focused primarily upon equilibrium molecular dynamics (MD) and Monte Carlo (MC) methods. These have been thoroughly reviewed by several investigators [70, 71].

Computational approach can play an important role in the development of the CNT-based composites by providing simulation results to help on the understanding, analysis and design of such nanocomposites. At the nanoscale, analytical models are difficult to establish or too complicated to solve, and tests are extremely difficult and expensive to conduct. Modeling and simulations of nanocomposites, on the other hand, can be achieved readily and cost effectively on even a desktop computer. Characterizing the mechanical properties of CNT-based composites is just one of the many important and urgent tasks that simulations can follow out [72].

Computer simulations on model systems have in recent years provided much valuable information on the thermodynamic, structural and transport properties of classical dense fluids. The success of these methods rests primarily on the fact that a model containing a relatively small number of particles is in general found to be sufficient to simulate the behavior of a macroscopic system. Two distinct techniques of computer simulation have been developed which are known as the method of molecular dynamics and the Monte Carlo method [73–75].

Instead of adopting a trial-and-error approach to membrane development, it is far more efficient to have a real understanding of the separation phenomena to guide membrane design [76–79]. Similarly, methods such as MC, MD and other computational techniques have improved the understanding of the relationships between membrane characteristics and separation properties. In addition to these inputs, it is also beneficial to have simple models and theories that give an overall insight into separation performance [80–83].

11.5.1   PERMEANCE AND SELECTIVITY OF SEPARATION MEMBRANES

A membrane separates one component from another on the basis of size, shape or chemical affinity. Two characteristics dictate membrane performance, permeability, that is the flux of the membrane, and selectivity or the membrane's preference to pass one species and not another [84].

A membrane can be defined as a selective barrier between two phases, the "selective" being inherent to a membrane or a membrane processes. The

membrane separation technology is proving to be one of the most significant unit operations. The technology inherits certain advantages over other
methods. These advantages include compactness and light weight, low labor
intensity, modular design that allows for easy expansion or operation at partial
capacity, low maintenance, low energy requirements, low cost, and environmentally friendly operations. A schematic representation of a simple separation membrane process is shown in Fig. 11.16.

A feed stream of mixed components enters a membrane unit where it is
separated into a retentate and permeate stream. The retentate stream is typically the purified product stream and the permeate stream contains the waste
component.

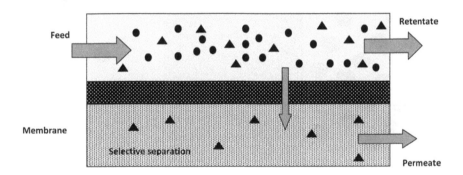

FIGURE 11.16   Schematic of membrane separation.

A quantitative measure of transport is the flux (or permeation rate), which
is defined as the number of molecules that pass through a unit area per unit
time [85]. It is believed that this molecular flux follows Fick's first law. The
flux is proportional to the concentration gradient through the membrane.
There is a movement from regions of high concentration to regions of low
concentration, which may be expressed in the form:

$$J = -D\frac{dc}{dx} \tag{5}$$

By assuming a linear concentration gradient across the membrane, the flux
can be approximated as:

$$J = -D\frac{C_2 - C_1}{L} \tag{6}$$

where $C_1 = c(0)$ and $C_2 = c(L)$ are the downstream and upstream concentrations (corresponding to the pressures p_1 and p_2 via sorption isotherm $c(p)$, respectively, and L is the membrane thickness, as labeled in Fig. 11.17.

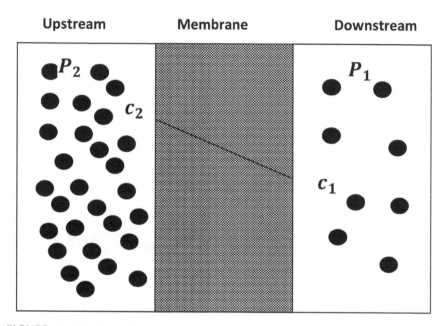

FIGURE 11.17   Separation membrane with a constant concentration gradient across membrane thickness L.

The membrane performance of various materials is commonly compared using the thickness, independent material property, and the permeability, which is related to the flux as:

$$P = \frac{JL}{P_2 - P_1} = \left(\frac{C_2 - C_1}{P_2 - P_1} \right) D \tag{7}$$

In the case where the upstream pressure is much greater than the downstream pressure ($p_2 >> p_1$ and $C_2 >> C_1$) the permeability can be simplified so:

$$P = \frac{C_2}{P_2} D \tag{8}$$

The permeability is more commonly used to describe the performance of a membrane than flux. This is because the permeability of a homogenous stable membrane material is constant regardless of the pressure differential or membrane thickness and hence it is easier to compare membranes made from different materials.

By introducing a solubility coefficient, the ratio of concentration over pressure C_2/p_2, when sorption isotherm can be represented by the Henry's law, the permeability coefficient may be expressed simply as:

$$P = SD \qquad (9)$$

This form is useful as it facilitates the understanding of this physical property by representing it in terms of two components:

Solubility which is an equilibrium component describing the concentration of gas molecules within the membrane, that is the driving force, and

Diffusivity, which is a dynamic component describing the mobility of the gas molecules within the membrane.

The separation of a mixture of molecules A and B is characterized by the selectivity or ideal separation factor $\alpha_{A/B} = P(A)/P(B)$, the ratio of permeability of the molecule A over the permeability of the molecule B. According to Eq. 9, it is possible to make separations by diffusivity selectivity $D(A)/D(B)$ or solubility selectivity $S(A)/S(B)$ [85–86]. This formalism is known in membrane science as the solution-diffusion mechanism. Since the limiting stage of the mass transfer is overcoming of the diffusion energy barrier, this mechanism implies the activated diffusion. Because of this, the temperature dependences of the diffusion coefficients and permeability coefficients are described by the Arrhenius equations.

Gas molecules that encounter geometric constrictions experience an energy barrier such that sufficient kinetic energy of the diffusing molecule or the groups that form this barrier, in the membrane is required in order to overcome the barrier and make a successful diffusive jump. The common form of the Arrhenius dependence for the diffusion coefficient can be expressed as:

$$D_A = D_A^* \exp(-\Delta E_a / RT) \qquad (10)$$

For the solubility coefficient the Van't Hoff equation holds:

$$S_A = S_A^* \exp(-\Delta H_a / RT) \qquad (11)$$

Where $\Delta H_a < 0$ is the enthalpy of sorption. From Eq. 9, it can be written:

$$P_A = P_A^* \exp(-\Delta E_p / RT) \qquad (12)$$

where $\Delta E_p = \Delta E_a + \Delta H_a$ are known to diffuse within nonporous or porous membranes according to various transport mechanisms. Table 11.3 illustrates the mechanism of transport depending on the size of pores. For very narrow pores, size-sieving mechanism is realized that can be considered as a case of activated diffusion. This mechanism of diffusion is most common in the case of extensively studied nonporous polymeric membranes. For wider pores, the surface diffusion (also an activated diffusion process) and the Knudsen diffusion are observed [87–89].

TABLE 11.3   Transport Mechanisms

Mechanism	Schamatic	Process
Activated diffusion		Constriction energy barrier
Surface diffusion		Adsorption – site energy barrier
Knudsen diffusion		Direction and velocity

Sorption does not necessarily follow Henry's law. For a glassy polymer an assumption is made that there are small cavities in the polymer and the sorption at the cavities follows Langmuir's law. Then, the concentration in the membrane is given as the sum of Henry's law adsorption and Langmuir's law adsorption

$$C = K_P\, P + \frac{C_h^* \, b_P}{1 + b_P} \qquad (13)$$

It should be noted that the applicability of solution (sorption)-diffusion model has nothing to do with the presence or absence of the pore.

11.5.2 DIFFUSIVITY

The diffusivity through membranes can be calculated using the time-lag method [90]. A plot of the flow through the membrane versus time reveals an initial transient permeation followed by steady state permeation. Extending the linear section of the plot back to the intersection of the x-axis gives the value of the time-lag (θ) as shown in Fig. 11.18.

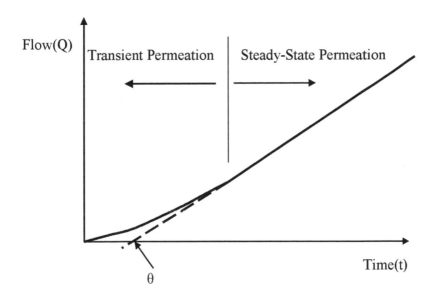

FIGURE 11.18 Calculation of the diffusion coefficient using the time-lag method, once the gradient is constant and steady state flow through the membrane has been reached, a extrapolation of the steady state flow line back to the x-axis where the flow is 0 reveals the value of the time lag (θ).

The time lag relates to the time it takes for the first molecules to travel through the membrane and is thus related to the diffusivity. The diffusion coefficient can be calculated from the time-lag and the membrane thickness as shown in Equation (14) [91–92].

$$D = \frac{\Delta x}{6\theta} \qquad (14)$$

Surface diffusion is the diffusion mechanism which dominates in the pore size region between activation diffusion and Knudsen diffusion [93].

11.5.3 SURFACE DIFFUSION

A model that well described the surface diffusion on the pore walls was proposed many years ago. It was shown to be consistent with transport parameters in porous polymeric membranes. When the pore size decreased below a certain level, which depends on both membrane material and the permeability coefficient exceeds the value for free molecular flow (Knudsen diffusion), especially in the case of organic vapors. Note that surface diffusion usually occurs simultaneously with Knudsen diffusion but it is the dominant mechanism within a certain pore size. Since surface diffusion is also a form of activated diffusion, the energy barrier is the energy required for the molecule to jump from one adsorption site to another across the surface of the pore. By allowing the energy barrier to be proportionate to the enthalpy of adsorption, Gilliland et al., [94] established an equation for the surface diffusion coefficient expressed here as:

$$D_S = D_S^* \exp(\frac{-aq}{RT}) \qquad (15)$$

Where is a preexponential factor depending on the frequency of vibration of the adsorbed molecule normal to the surface and the distance from one adsorption site to the next. The quantity the heat of adsorption is $(q > 0)$ and a proportionality constant is $(0 < a < 1)$. The energy barrier separates the adjacent adsorption sites. An important observation is that more strongly adsorbed molecules are less mobile than weakly adsorbed molecules [95].

In the case of surface diffusion, the concentration is well described by Henry's law $c = Kp$, where K is $K = K_0 \exp(q/RT)$ [95, 96]. Since solubility is the ratio of the equilibrium concentration over pressure, the solubility is equivalent to the Henry's law coefficient.

$$S_S = K_0 \exp(q|RT) \qquad (16)$$

Which implies the solubility is a decreasing function of temperature. The product of diffusivity and solubility gives:

$$P_S = P_S^* \exp\left(\frac{(1-a)q}{RT}\right) \tag{17}$$

Since $0 < a < 1$ the total permeability will decrease with increased temperature meaning that any increase in the diffusivity is counteracted by a decrease in surface concentration [95].

11.5.4 KNUDSEN DIFFUSION

Knudsen diffusion [95, 97–99] depending on pressure and mean free path which applies to pores between 10 Å and 500 Å in size [100]. In this region, the mean free path of molecules is much larger than the pore diameter. It is common to use Knudsen number $K_n = \lambda/d$ to characterize the regime of permeation through pores. When $K_n \ll 1$, viscous (Poiseuille) flow is realized. The condition for Knudsen diffusion is $Kn \gg 1$. An intermediate regime is realized when $K_n \approx 1$. The Knudsen diffusion coefficient can be expressed in the following form:

$$D_K = \frac{d}{3\tau}\bar{u} \tag{18}$$

This expression shows that the separation outcome should depend on the differences in molecular speed (or molecular mass). The average molecular speed is calculated using the Maxwell speed distribution as:

$$\bar{u} = \sqrt{\frac{8RT}{\pi m}} \tag{19}$$

And the diffusion coefficient can be presented as:

$$D_K = \left(\frac{d}{3\tau}\right)(\frac{8RT}{\pi m} \tag{20}$$

For the flux in the Knudsen regime the following equation holds [101, 102]:

$$J = n\pi d^2 \Delta p D_K / 4RTL \tag{21}$$

After substituting Eq. (20) into Eq. (21), one has the following expressions for the flux and permeability coefficient is:

$$ J = \left(\frac{n\pi^{\frac{1}{2}}d^3 \Delta p}{6\tau L} \right)(\frac{2}{mRT})^{1/2} \tag{22} $$

$$ P = \left(\frac{n\pi^{\frac{1}{2}}d^3}{6\tau} \right)(\frac{2}{mRT})^{1/2} \tag{23} $$

Two important conclusions can be made from analysis of Eqs. (22) and (23). First, selectivity of separation in Knudsen regime is characterized by the ratio $\alpha_{ij} = (M_j/M_i)^{1/2}$. It means that membranes where Knudsen diffusion predominates are poorly selective.

The most common approach to obtain diffusion coefficients is equilibrium molecular dynamics. The diffusion coefficient that is obtained is a self-diffusion coefficient. Transport-related diffusion coefficients are less frequently studied by simulation but several approaches using nonequilibrium MD (NEMD) simulation can be used.

11.5.5   MOLECULAR DYNAMICS (MD) SIMULATIONS

Conducting experiments for material characterization of the nano composites is a very time consuming, expensive and difficult. Many researchers are now concentrating on developing both analytical and computational simulations. MD simulations are widely being used in modeling and solving problems based on quantum mechanics. Using Molecular dynamics it is possible to study the reactions, load transfer between atoms and molecules. If the objective of the simulation is to study the overall behavior of CNT-based composites and structures, such as deformations, load and heat transfer mechanisms then the continuum mechanics approach can be applied safely to study the problem effectively [103].

MD tracks the temporal evolution of a microscopic model system by integrating the equations of motion for all microscopic degrees of freedom. Numerical integration algorithms for initial value problems are used for this purpose, and their strengths and weaknesses have been discussed in simulation texts [104–106].

MD is a computational technique in which a time evolution of a set of interacting atoms is followed by integrating their equations of motion. The forces between atoms are due to the interactions with the other atoms. A trajectory is calculated in a 6-N dimensional phase space (three position and three momentum components for each of the N atoms). Typical MD simulations of CNT composites are performed on molecular systems containing up to tens of thousands of atoms and for simulation times up to nanoseconds. The physical quantities of the system are represented by averages over configurations distributed according to the chosen statistical ensemble. A trajectory obtained with MD provides such a set of configurations. Therefore the computation of a physical quantity is obtained as an arithmetic average of the instantaneous values. Statistical mechanics is the link between the nanometer behavior and thermodynamics. Thus the atomic system is expected to behave differently for different pressures and temperatures [107].

The interactions of the particular atom types are described by the total potential energy of the system, U, as a function of the positions of the individual atoms at a particular instant in time

$$U = U\left(X_i, \ldots\ldots, X_n\right) \qquad (24)$$

where $_1$ represents the coordinates of atom i in a system of N atoms. The potential equation is invariant to the coordinate transformations, and is expressed in terms of the relative positions of the atoms with respect to each other, rather than from absolute coordinates [107].

MD is readily applicable to a wide range of models, with and without constraints. It has been extended from the original micro canonical ensemble formulation to a variety of statistical mechanical ensembles. It is flexible and valuable for extracting dynamical information. The Achilles' heel of MD is its high demand of computer time, as a result of which the longest times that can be simulated with MD fall short of the longest relaxation times of most real-life macromolecular systems by several orders of magnitude. This has two important consequences .(a) Equilibrating an atomistic model polymer system with MD alone is problematic; if one starts from an improbable configuration, the simulation will not have the time to depart significantly from that configuration and visit the regions of phase space that contribute most significantly to the properties. (b) Dynamical processes with characteristic times longer than approximately 10^{-7} s cannot be probed directly; the relevant correlation functions do not decay to zero within the simulation time and thus their long-time tails are in accessible, unless some extrapolation is invoked based on their short-time behavior.

Recently, rigorous multiple time step algorithms have been invented, which can significantly augment the ratio of simulated time to CPU time.

Such an algorithm is the reversible Reference System Propagator Algorithm (rRESPA) [108–109]. This algorithm invokes a Trotter factorization of the Liouville operator in the numerical integration of the equations of motion: fast-varying (e.g., bond stretching and bond angle bending) forces are updated with a short time step , while slowly varying forces (e.g., non bonded interactions, which are typically expensive to calculate, are updated with a longer time step . Usingand , one can simulate 300 ns of real time of a polyethylene melt on a modest workstation [110]. This is sufficient for the full relaxation of a system of C_{250} chains, but not of longer-chain systems.

A paper of Furukawa and Nitta is cited first to understand the NEMD simulation semiquantitatively, since, even though the paper deals with various pore shapes, complicated simulation procedure is described clearly.

MD simulation is more preferable to study the nonequilibrium transport properties. Recently some NEMD methods have also been developed, such as the grand canonical molecular dynamics (GCMD) method [111, 112] and the dual control volume GCMD technique (DCV-GCMD) [113, 114].These methods provide a valuable clue to insight into the transport and separation of fluids through a porous medium. The GCMD method has recently been used to investigate pressure-driven and chemical potential-driven gas transport through porous inorganic membrane [115].

11.5.5.1   EQUILIBRIUM MD SIMULATION

A self-diffusion coefficient can be obtained from the mean-square displacement (MSD) of one molecule by means of the Einstein equation in the form [115]:

$$D_A^* = \frac{1}{6N_\alpha} \lim_{t \to \infty} \frac{d}{dt} \left(r_i(t) - r_i(0) \right)^2 \qquad (25)$$

Where Na is the number of molecules, and are the initial and final (at time t) positions of the center of mass of one molecule i over the time interval t, and is MSD averaged over the ensemble. The Einstein relationship assumes a random walk for the diffusing species. For slow diffusing species, anomalous diffusion is sometimes observed and is characterized by:

$$\left(r_i(t) - r_i(0) \right)^2 \propto t^n \qquad (26)$$

where n < 1 (n = 1 for the Einstein diffusion regime). At very short times (t < 1 ps), the MSD may be quadratic iv n time (n = 2) which is characteristic of 'free flight' as may occur in a pore or solvent cage prior to collision with the pore or cage wall. The result of anomalous diffusion, which may or may not occur in intermediate time scales, is to create a smaller slope at short times, resulting in a larger value for the diffusion coefficient. At sufficiently long times (the hydrodynamic limit), a transition from anomalous to Einstein diffusion (n = 1) may be observed [71].

An alternative approach to MSD analysis makes use of the center-of-mass velocity autocorrelation function (VACF) or Green–Kubo relation, given as follows [116]:

$$D = \frac{1}{3}\int (v_i(t).v_i(0))dt \qquad (27)$$

Concentration in the simulation cell is extremely low and its diffusion coefficient is an order of magnitude larger than that of the polymeric segments. Under these circumstances, the self-diffusion and mutual diffusion coefficients of the penetrant are approximately equal, as related by the Darken equation in the following form:

$$D_{AB} = (D_A^* x_B + D_B^* x_A)\left(\frac{d\ln f_A}{d\ln c_A}\right) \qquad (28)$$

In the limit of low concentration of diffusion , Eq. (28) reduces to:

$$D_A^* \equiv D_{AB} \qquad (29)$$

11.5.5.2 NON-EQUILIBRIUM MD SIMULATION

Experimental diffusion coefficients, as obtained from time-lag measurements, report a transport diffusion coefficient which cannot be obtained from equilibrium MD simulation. Comparisons made in the simulation literature are typically between time-lag diffusion coefficients (even calculated for glassy polymers without correction for dual-mode contributions and self-diffusion coefficients. As discussed above, mutual diffusion coefficients can be obtained directly from equilibrium MD simulation but simulation of transport diffusion coefficients require the use of NEMD methods, that are less commonly available and more computationally expensive [117].

For these reasons, they have not been frequently used. One successful approach is to simulate a chemical potential gradient and combine MD with GCMC methods (GCMC–MD), as developed by Hoeffel Finger et al., [114] and Mac Elroy [118]. This approach has been used to simulate permeation of a variety of small molecules through nanoporous carbon membranes, carbon nanotubes, porous silica and self-assembled mono layers [119–121]. A diffusion coefficient then can be obtained from the relation:

$$D = \frac{KT}{F}(V) \qquad (30)$$

11.5.6   GRAND CANONICAL MONTE CARLO (GCMC) SIMULATION

A standard GCMC simulation is employed in the equilibrium study, while MD simulation is more preferable to study the nonequilibrium transport properties [104].

Monte Carlo method is formally defined by the following quote as: Numerical methods that are known as Monte Carlo methods can be loosely described as statistical simulation methods, where statistical simulation is defined in quite general terms to be any method that uses sequences of random numbers to perform the simulation [122].

The name "Monte Carlo" was chosen because of the extensive use of random numbers in the calculations [104].One of the better known applications of Monte Carlo simulations consists of the evaluation of integrals by generating suitable random numbers that will fall within the area of integration. A simple example of how a MC simulation method is applied to evaluate the value of π is illustrated in Fig. 11.19. By considering a square that inscribes a circleof a diameter R, one can deduce that the area of the square is R^2, and the circle has anarea of $\pi R^2/4$.Thus, the relative area of the circle and the square will be $\pi/4$. A large number of two independent random numbers (with x and y coordinates) of trial shots is generated within the square to determine whether each of them falls inside of the circle or not. After thousands or millions of trial shots, the computer program keeps counting the total number of trial shots inside the square and the number of shots landing inside the circle. Finally, the value of $\pi/4$ can be approximated based on the ratio of the number of shots that fall inside the circle to the total number of trial shots.

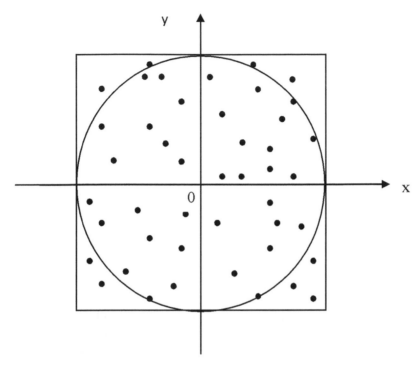

FIGURE 11.19 Illustration of the application of the Monte Carlo simulation method for the calculation of the value of π by generating a number of trial shots, in which the ratio of the number of shots inside the circle to the total number of trial shots will approximately approach the ratio of the area of the circle to the area of the square.

As stated earlier, the value of an integral can be calculated via MC methods by generating a large number of random points in the domain of that integral. Equation (31) shows a definite integral:

$$F = \int_a^b f(x)\,dx \tag{31}$$

Where $f(x)$ is a continuous and real-valued function in the interval $[a, b]$. The integral can be rewritten as [104]:

$$F = \int_a^b dx \left(\frac{f(x)}{\rho(x)} \right) \rho(x) \cong \frac{f(\xi_i)}{\rho(\xi_i)}_\tau \tag{32}$$

If the probability function is chosen to be a continuous uniform distribution, then:

$$\rho(x) = \frac{1}{(b-a)} \qquad a \le x \le b \qquad (33)$$

Subsequently, the integral, F, can be approximated as:

$$F \approx \frac{(b-a)}{\tau} \sum_{i=1}^{\tau} f(\xi_i) \qquad (34)$$

In a similar way to the MC integration methods, MC molecular simulation methods rely on the fact that a physical system can be defined to possess a definite energy distribution function, which can be used to calculate thermodynamic properties.

The applications of MC are diverse such as Nuclear reactor simulation, Quantum chromo dynamics, Radiation cancer therapy, Traffic flow, Stellar evolution, Econometrics, Dow Jones forecasting, Oil well exploration, VSLI design [122]

The MC procedure requires the generation of a series of configurations of the particles of the model in a way which ensures that the configurations are distributed in phase space according to some prescribed probability density.

The mean value of any configurational property determined from a sufficiently large number of configurations provides an estimate of the ensemble-average value of that quantity; the nature of the ensemble average depends upon the chosen probability density.

These machine calculations provide what is essentially exact information on the consequences of a given intermolecular force law. Application has been made to hard spheres and hard disks, to particles interacting through a Lenard-Jones 12–6 potential function and other continuous potentials of interest in the study of simple fluids, and to systems of charged particles [123].

The MC technique is a stochastic simulation method designed to generate a long sequence, or 'Markov chain' of configurations that asymptotically sample the probability density of an equilibrium ensemble of statistical mechanics [105, 116]. For example, a MC simulation in the canonical (NVT) ensemble, carried out under the macroscopic constraints of a prescribed number of molecules N, total volume V and temperature T, samples configurations r_p with probability proportional to , with, k_B being the Boltzmann constant and T the absolute temperature. Thermodynamic properties are computed as averages over all sampled configurations.

The efficiency of a MC algorithm depends on the elementary moves it employs to go from one configuration to the next in the sequence. An attempted move typically involves changing a small number of degrees of freedom; it is accepted or rejected according to selection criteria designed so that the sequence ultimately conforms to the probability distribution of interest. In addition to usual moves of molecule translation and rotation practiced for small-molecule fluids, special moves have been invented for polymers. The reptation (slithering snake) move for polymer chains involves deleting a terminal segment on one end of the chain and appending a terminal segment on the other end, with the newly created torsion angle being assigned a randomly chosen value [124].

In most MC algorithms the overall probability of transition from some state (configuration) m to some other state n, as dictated by both the attempt and the selection stages of the moves, equals the overall probability of transition from n to m; this is the principle of detailed balance or 'microscopic reversibility.' The probability of attempting a move from state m to state n may or may not be equal to that of attempting the inverse move from state n to state m. These probabilities of attempt are typically unequal in 'bias' MC algorithms, which incorporate information about the system energetics in attempting moves. In bias MC, detailed balance is ensured by appropriate design of the selection criterion, which must remove the bias inherent in the attempt [105, 116].

11.5.7 MEMBRANE MODEL AND SIMULATION BOX

The MD simulations [125] can be applied for the permeation of pure and mixed gasses across carbon membranes with three different pore shapes: the diamond pore (DP), zigzag path (ZP) and straight path (SP), each composed of micrographite crystalline. Three different pore shapes can be considered: DP, ZP and SP.

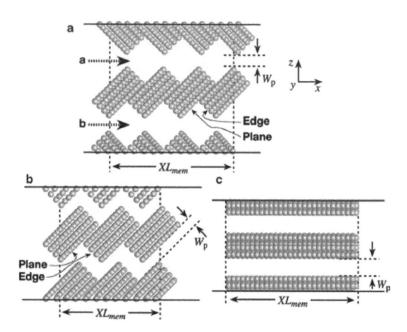

FIGURE 11.20   Three membrane pore shapes; (a) diamond path (DP), (b) zigzag path (ZP), (c) straight path (SP).

Figure 11.20. (a)–(c)shows the cross-sectional view of each pore shape. DP (A) has two different pore mouths; one a large (pore a) and the other a small mouth (pore b). ZP (B) has zigzag shaped pores whose sizes (diameters) are all the same at the pore entry. SP (C) has straight pores which can be called slit-shaped pores.

In a simulation system, we investigate the equilibrium selective adsorption and nonequilibrium transport and separation of gas mixture in the nanoporous carbon membrane are modeled as slits from the layer structure of graphite. A schematic representation of the system used in our simulations is shown in Fig. 11.21(a) and (b), in which the origin of the coordinates is at the center of simulation box and transport takes place along the x-direction in the nonequilibrium simulations. In the equilibrium simulations, the box as shown in Fig. 11.21(a) is employed, whose size is set as 85.20 nm × 4.92 nm × (1.675 + W) nm in x-, y-, and z-directions, respectively, where W is the pore width, i.e. the separation distance between the centers of carbon atoms on the two layers forming a slit pore (Fig. 11.21). L_{cc} is the separation distance between two centers of adjacent carbon atom; L_m is the pore length; W is the pore width, Δ is the separation distance between two carbon atom centers of two adjacent layers [126].

The simulation box is divided into three regions where the chemical potential for each component is the same. The middle region (M-region) represents the membrane with slit pores in which the distances between the two adjacent carbon atoms (Lcc) and two adjacent graphite basal planes (Δ).

Period boundary conditions are employed in all three directions. In the nonequilibrium molecular dynamics simulations in order to use period boundary conditions in three directions, we have to divide the system into five regions as shown in Fig. 11.21(b). (Fig. 11.22)

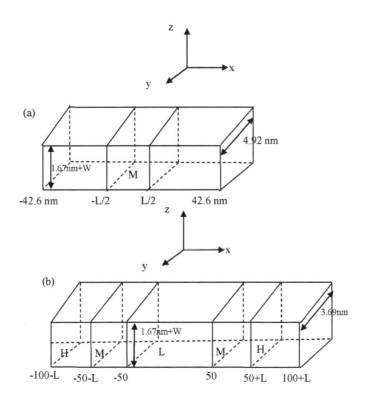

FIGURE 11.21 Schematic representation of the simulation boxes. The H-, L-and M-areas correspond to the high and low chemical potential control volumes, and membrane, respectively. Transport takes place along the x-direction in the nonequilibrium simulations. (a) Equilibrium adsorption simulations and (b) nonequilibrium transport simulations. L is the membrane thickness and W is the pore width.

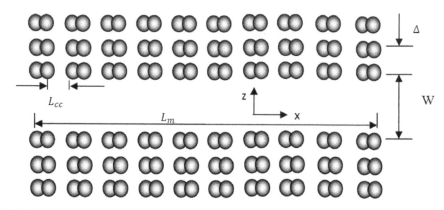

FIGURE 11.22   Schematic representation of slit pore.

Each symmetric box has three regions. Two are density control; H-region (high density) and L-region (low density) and one is free of control M-region which is placed between the H-and L-region. For each simulation, the density in the H-region, ρ_H, is maintained to be that of the feed gas and the density in the L-region is maintained at zero, corresponding to the vacuum.

The difference in the gas density between the H-and L-region is the driving force for the gas permeation through the M-region which represents the membrane.

The transition and rotational velocities are given to each inserted molecules randomly based on the Gaussian distribution around an average velocity corresponding to the specified temperature.

Molecules spontaneously move from H-to L-region via leap-frog algorithm and a nonequilibrium steady state is obtained at the M-region. During a simulation run, equilibrium with the bulk mass at the feed side at the specified pressure and temperature is maintained at the H-region by carrying out GCMC creations and destructions in terms of the usual acceptance criteria [28]. Molecules entered the L-region were moved out immediately to keep vacuum. The velocities of newly inserted molecules were set to certain values in terms of the specified temperature by use of random numbers on the Gaussian distribution.

11.6 CONCLUSION

The concept of membrane processes is relatively simple but nevertheless often unknown. Membrane separation processes can be used for a wide range of applications. The separation mechanism in MF/UF/NF is mainly the size exclusion, which is indicated in the nominal ratings of the membranes. The other separation mechanism includes the electrostatic interactions between solutes and membranes, which depends on the surface and physiochemical properties of solutes and membranes. The available range of membrane materials includes polymeric, carbon, silica, zeolite and other ceramics, as well as composites. Each type of membrane can have a different porous structure. Nowadays, there are more reports on the fluid transport through porous CNTs/polymer membrane. Computational approach can play an important role in the development of the CNT-based composites by providing simulation results to help on the understanding, analysis and design of such nanocomposites. Computational approaches to obtain solubility and diffusion coefficients of small molecules in polymers have focused primarily upon molecular dynamics and Monte Carlo methods. Molecular dynamics simulations are widely being used in modeling and solving problems based on quantum mechanics. Using Molecular dynamics it is possible to study the reactions, load transfer between atoms and molecules. Monte Carlo molecular simulation methods rely on the fact that a physical system can be defined to possess a definite energy distribution function, which can be used to calculate thermodynamic properties. The Monte Carlo technique is a stochastic simulation method designed to generate a long sequence, or 'Markov chain' of configurations that asymptotically sample the probability density of an equilibrium ensemble of statistical mechanics. So using from molecular dynamic or Monte Carlo techniques can be useful to simulate the membrane separation process depends on the purpose and the condition of process.

KEYWORDS

- **Computational methods**
- **Filtration**
- **Membrane**
- **Membrane types**

REFERENCES

1. Majeed, S., et al. (2012). Multi-Walled Carbon Nanotubes (MWCNTs) Mixed Polyacrylo-nitrile (PAN) Ultrafiltration Membranes. *Journal of Membrane Science*, 403, 101–109.
2. Macedonio, F. & Drioli, E. (2008). *Pressure-Driven Membrane Operations and Membrane Distillation Technology Integration for Water Purification.* Desalination, *223(1)*, 396–409.
3. Merdaw, A. A., Sharif, A. O. & Derwish, G. A. W. (2011). Mass Transfer in Pressure-Driven Membrane Separation Processes, Part II. *Chemical Engineering Journal, 168(1)*, 229–240.
4. Van Der Bruggen, B., et al. (2003). *A Review of Pressure-Driven Membrane Processes in Wastewater Treatment and Drinking Water Production.* Environmental Progress, *22(1)*, 46–56.
5. Cui, Z. F. & Muralidhara, H. S. (2010). *Membrane Technology: A Practical Guide to Membrane Technology and Applications in Food and Bioprocessing* Elsevier. 288.
6. Shirazi, S., Lin, C. J. & Chen, D. (2010). *Inorganic Fouling of Pressure-Driven Membrane Processes A Critical Review.* Desalination, *250(1)*, 236–248.
7. Pendergast, M. M. & Hoek, E. M. V. (2011). *A Review of Water Treatment Membrane Nanotechnologies.* Energy & Environmental Science, *4(6)*, 1946–1971.
8. Hilal, N., et al. (2004). *A comprehensive review of nanofiltration membranes: Treatment, pretreatment, modeling, and atomic force microscopy.* Desalination, *170(3)*, 281–308.
9. Srivastava, A., Srivastava, S. & Kalaga, K. (2013). *Carbon Nanotube Membrane Filters*, in *Springer Handbook of Nanomaterials* Springer, 1099–1116.
10. Colombo, L. & Fasolino, A. L. (2010). *Computer-Based Modeling of Novel Carbon Systems and Their Properties: Beyond Nanotubes, 3* Springer. 258.
11. Polarz, S. & Smarsly, B. (2002). Nanoporous Materials. *Journal of Nanoscience and Nanotechnology, 2(6)*, 581–612.
12. Gray-Weale, A. A., et al. (1997). *Transition-State Theory Model for the Diffusion Coefficients of Small Penetrants in Glassy Polymers.* Macromolecules, *30(23)*, 7296–7306.
13. Rigby, D. & Roe, R. (1987). Molecular Dynamics Simulation of Polymer Liquid and Glass. I. Glass Transition. *The Journal of Chemical Physics, 87*, 7285.
14. Freeman, B. D., Yampolskii, Y. P. & Pinnau, I. (2006). *Materials Science of Membranes for Gas and Vapor Separation* Wiley. com. 466.
15. Hofmann, D., et al. (2003). *Molecular Modeling Investigation of Free Volume Distributions in Stiff Chain Polymers with Conventional and Ultrahigh Free Volume: Comparison Between Molecular Modeling and Positron Lifetime Studies.* Macromolecules, *36(22)*, 8528–8538.
16. Greenfield, M. L. & Theodorou, D. N. (1993). *Geometric Analysis of Diffusion Pathways in Glassy and Melt Atactic Polypropylene.* Macromolecules, *26(20)*, 5461–5472.
17. Baker, R. W. (2012). *Membrane Technology and Applications.* John Wiley & Sons. 592
18. Strathmann, H., Giorno, L. & Drioli, E. (2011). *Introduction to Membrane Science and Technology* Wiley-VCH Verlag & Company. 544.
19. Chen, J. P., et al. (2008). *Membrane Separation: Basics and Applications*, in *Membrane and Desalination Technologies*, L. K. Wang, et al, Editors Humana Press, 271–332.
20. Mortazavi, S. (2008). *Application of Membrane Separation Technology to Mitigation of Mine Effluent and Acidic Drainage* Natural Resources Canada. 194.
21. Porter, M. C. (1990). *Handbook of Industrial Membrane Technology.* Noyes Publications. 604.

22. Naylor, T. V. (1996). *Polymer Membranes: Materials, Structures and Separation Performance.* Rapra Technology Limited. 136.
23. Freeman, B. D. (2012). *Introduction to Membrane Science and Technology. By Heinrich Strathmann.* Angewandte Chemie International Edition, *51(38)*, 9485–9485.
24. Kim, I., Yoon, H. & Lee, K. M. (2002). *Formation of Integrally Skinned Asymmetric Polyetherimide Nanofiltration Membranes by Phase Inversion Process.* Journal of Applied Polymer Science, *84(6)*, 1300–1307.
25. Khulbe, K. C., Feng, C. Y. & Matsuura, T. (2007). *Synthetic Polymeric Membranes: Characterization by Atomic Force Microscopy.* Springer. 198.
26. Loeb, L. B. (2004). *The Kinetic Theory of Gases* Courier Dover Publications. 678.
27. Koros, W. J. & Fleming, G. K. (1993). Membrane-Based Gas Separation. *Journal of Membrane Science, 83(1)*, 1–80.
28. Perry, J. D., Nagai, K. & Koros, W. J. (2006). *Polymer Membranes for Hydrogen Separations.* MRS Bulletin, *31(10)*, 745–749.
29. Yang, W., et al. (2007). *Carbon Nanotubes for Biological and Biomedical Applications.* Nanotechnology, *18(41)*, 412001.
30. Bianco, A., et al. (2005). *Biomedical Applications of Functionalized Carbon Nanotubes.* Chemical Communications, *5*, 571–577.
31. Salvetat, J., et al. (1999). *Mechanical Properties of Carbon Nanotubes.* Applied Physics A, *69(3)*, 255–260.
32. Zhang, X., et al. (2007). *Ultrastrong, Stiff, and Lightweight Carbon-Nanotube Fibers.* Advanced Materials, *19(23)*, 4198–4201.
33. Arroyo, M. & Belytschko, T. (2004). *Finite Crystal Elasticity of Carbon Nanotubes Based on the Exponential Cauchy-Born Rule.* Physical Review B, *69(11)*, 115415.
34. Wang, J., et al. (2006). *Energy and Mechanical Properties of Single-Walled Carbon Nanotubes Predicted Using the Higher Order Cauchy-Born rule.* Physical Review B, *73(11)*, 115428.
35. Zhang, Y. (2011). *Single-Walled Carbon Nanotube Modeling Based on One-and Two-Dimensional Cosserat Continua.* University of Nottingham.
36. Wang, S. (2006). *Functionalization of Carbon Nanotubes: Characterization, Modeling and Composite Applications.* Florida State University. 193.
37. Lau, K.-T., Gu, C. & Hui, D. (2006). *A Critical Review on Nanotube and Nanotube/ Nanoclay Related Polymer Composite Materials.* Composites Part B: Engineering, *37(6)*, 425–436.
38. Choi, W., et al. (2010). *Carbon Nanotube-Guided Thermopower Waves.* Materials Today, *13(10)*, 22–33.
39. Iijima, S. (1991). *Helical Microtubules of Graphitic Carbon.* Nature, *354(6348)*, 56–58.
40. Sholl, D. S. & Johnson, J. (2006). *Making High-Flux Membranes with Carbon Nanotubes.* Science, *312(5776)*, 1003–1004.
41. Zang, J., et al. (2009). *Self-Diffusion of Water and Simple Alcohols in Single-Walled Aluminosilicate Nanotubes.* ACS Nano, *3(6)*, 1548–1556.
42. Talapatra, S., Krungleviciute, V. & Migone, A. D. (2002). *Higher Coverage Gas Adsorption on the Surface of Carbon Nanotubes: Evidence for a Possible New Phase in the Second Layer.* Physical Review Letters, *89(24)*, 246106.
43. Pujari, S., et al. (2009). Orientation Dynamics in Multiwalled Carbon Nanotube Dispersions Under Shear Flow. *The Journal of chemical physics*, 130, 214903.
44. Singh, S. & Kruse, P. (2008). Carbon Nanotube Surface Science. *International Journal of Nanotechnology, 5(9)*, 900–929.

45. Baker, R. W. (2002). *Future Directions of Membrane Gas Separation Technology.* Industrial & Engineering Chemistry Research, *41(6)*, 1393–1411.
46. Erucar, I. & Keskin, S. (2011). *Screening Metal–Organic Framework-Based Mixed-Matrix Membranes for CO_2/CH_4 Separations.* Industrial & Engineering Chemistry Research, *50(22)*, 12606–12616.
47. Bethune, D. S., et al. (1993). *Cobalt-Catalyzed Growth of Carbon Nanotubes with Single-Atomic-Layer Walls.* Nature, *363*, 605–607.
48. Iijima, S. & Ichihashi, T. (1993). *Single-Shell Carbon Nanotubes of 1-nm Diameter.* Nature, *363*, 603–605.
49. Treacy, M., Ebbesen, T. & Gibson, J. (1996). *Exceptionally high Young's modulus observed for individual carbon nanotubes.*
50. Wong, E. W., Sheehan, P. E. & Lieber, C. (1997). *Nanobeam Mechanics: Elasticity, Strength, and Toughness of Nanorods and Nanotubes.* Science, *277(5334)*, 1971–1975.
51. Thostenson, E. T., Li, C. & Chou, T. W. (2005). *Nanocomposites in Context.* Composites Science and Technology, *65(3)*, 491–516.
52. Barski, M., P. Kędziora, & M. Chwał. (2013). *Carbon Nanotube/Polymer Nanocomposites: A Brief Modeling Overview.* Key Engineering Materials, 542, 29–42.
53. Dresselhaus, M. S., Dresselhaus, G. & Eklund, P. C. (1996). *Science of Fullerenes and Carbon Nanotubes: Their Properties and Applications.* Academic Press. 965.
54. Yakobson, B. & Smalley, R. E. (1997). *Some Unusual New Molecules—Long, Hollow Fibers with Tantalizing Electronic and Mechanical Properties—Have Joined Diamonds and Graphite in the Carbon Family.* Am Scientist, 85, 324–337.
55. Guo, Y. & Guo, W. (2003). Mechanical and Electrostatic Properties of Carbon Nanotubes under Tensile Loading and Electric Field. *Journal of Physics D: Applied Physics, 36(7)*, 805.
56. Berger, C., et al. (2006). *Electronic Confinement and Coherence in Patterned Epitaxial Graphene.* Science, *312(5777)*, 1191–1196.
57. Song, K., et al. (2013). *Structural Polymer-Based Carbon Nanotube Composite Fibers: Understanding the Processing–Structure–Performance Relationship.* Materials, *6(6)*, 2543–2577.
58. Park, O. K., et al. (2010). *Effect of Surface Treatment with Potassium Persulfate on Dispersion Stability of Multi-Walled Carbon Nanotubes.* Materials Letters, *64(6)*, 718–721.
59. Banerjee, S., T. Hemraj-Benny, & Wong, S. S. (2005). *Covalent Surface Chemistry of Single-Walled Carbon Nanotubes.* Advanced Materials, *17(1)*, 17–29.
60. Balasubramanian, K. & Burghard, M. (2005). *Chemically Functionalized Carbon Nanotubes.* Small, *1(2)*, 180–192.
61. Xu, Z. L. & F. Alsalhy Qusay. (2004). Polyethersulfone (PES) Hollow Fiber Ultrafiltration Membranes Prepared by PES/non-Solvent/NMP Solution. *Journal of Membrane Science, 233(1–2)*, 101–111.
62. Chung, T. S., Qin, J. J. & Gu, J. (2000). *Effect of Shear Rate Within the Spinneret on Morphology, Separation Performance and Mechanical Properties of Ultrafiltration Polyethersulfone Hollow Fiber Membranes.* Chemical Engineering Science, *55(6)*, 1077–1091.
63. Choi, J. H., Jegal, J. & Kim, W. N. (2007). *Modification of Performances of Various Membranes Using MWNTs as a Modifier.* Macromolecular Symposia, *249–250(1)*, 610–617.
64. Wang, Z. & Ma, J. (2012). *The Role of Nonsolvent in-Diffusion Velocity in Determining Polymeric Membrane Morphology.* Desalination, *286(0)*, 69–79.
65. Vilatela, J. J., Khare, R. & Windle, A. H. (2012). *The Hierarchical Structure and Properties of Multifunctional Carbon Nanotube Fiber Composites.* Carbon, *50(3)*, 1227–1234.

66. Benavides, R. E., Jana, S. C. & Reneker, D. H. (2012). *Nanofibers from Scalable Gas Jet Process.* ACS Macro Letters, *1(8)*, 1032–1036.
67. Gupta, V. B. & Kothari, V. K. (1997). *Manufactured Fiber Technology* Springer. 661.
68. Wang, T. & Kumar, S. (2006). Electrospinning of Polyacrylonitrile Nanofibers. *Journal of applied polymer science, 102(2)*, 1023–1029.
69. Song, K., et al. (2013). Lubrication of Poly (vinyl alcohol) Chain Orientation by Carbon Nano-Chips in Composite Tapes. *Journal of Applied Polymer Science, 127(4)*, 2977–2982.
70. Theodorou, D. N. (1996). *Molecular Simulations of Sorption and Diffusion in Amorphous Polymers.* Plastics Engineering-New York, 32, 67–142.
71. Müller-Plathe, F. (1994). *Permeation of polymers—a computational approach.* Acta Polymerica, *45(4)*, 259–293.
72. Liu, Y. J. & Chen, X. L. (2003). *Evaluations of the Effective Material Properties of Carbon Nanotube-Based Composites Using a Nanoscale Representative Volume Element.* Mechanics of materials, *35(1)*, 69–81.
73. Gusev, A. A. & Suter, U. W. (1993). Dynamics of Small Molecules in Dense Polymers Subject to Thermal Motion. *The Journal of chemical physics*, 99, 2228.
74. Elliott, J. A. (2011). *Novel Approaches to Multiscale Modeling in Materials Science.* International Materials Reviews, *56(4)*, 207–225.
75. Greenfield, M. L. & Theodorou, D. N. (1998). *Molecular Modeling of Methane Diffusion in Glassy Atactic Polypropylene via Multidimensional Transition State Theory.* Macromolecules, *31(20)*, 7068–7090.
76. Peng, F., et al. (2005). *Hybrid Organic-Inorganic Membrane: Solving the Tradeoff Between Permeability and Selectivity.* Chemistry of Materials, *17(26)*, 6790–6796.
77. Duke, M. C., et al. (2008). *Exposing the Molecular Sieving Architecture of Amorphous Silica Using Positron Annihilation Spectroscopy.* Advanced Functional Materials, *18(23)*, 3818–3826.
78. Hedstrom, J. A., et al. (2004). *Pore Morphologies in Disordered Nanoporous Thin Films.* Langmuir, *20(5)*, 1535–1538.
79. Pujari, P. K., et al. (2007). *Study of Pore Structure in Grafted Polymer Membranes Using Slow Positron Beam and Small-Angle X-ray Scattering Techniques.* Nuclear Instruments and Methods in Physics Research Section B: Beam Interactions with Materials and Atoms, *254(2)*, 278–282.
80. Wang, X. Y., et al. (2004). *Cavity Size Distributions in High Free Volume Glassy Polymers by Molecular Simulation.* Polymer, *45(11)*, 3907–3912.
81. Skoulidas, A. I. & Sholl, D. S. (2005). Self-Diffusion and Transport Diffusion of Light Gases in Metal-Organic Framework Materials Assessed Using Molecular Dynamics Simulations. *The Journal of Physical Chemistry B, 109(33)*, 15760–15768.
82. Wang, X. Y., et al. (2005). *A Molecular Simulation Study of Cavity Size Distributions and Diffusion in Para and Meta Isomers.* Polymer, *46(21)*, 9155–9161.
83. Zhou, J., et al. (2006). *Molecular Dynamics Simulation of Diffusion of Gases in Pure and Silica-Filled Poly (1-trimethylsilyl-1-propyne)[PTMSP].* Polymer, *47(14)*, 5206–5212.
84. Scholes, C. A., Kentish, S. E. & Stevens, G. W. (2008). *Carbon Dioxide Separation Through Polymeric Membrane Systems for Flue Gas Applications.* Recent Patents on Chemical Engineering, *1(1)*, 52–66.
85. Wijmans, J. G. & Baker, R. W. (2006). *The Solution-Diffusion Model: A Unified Approach to Membrane Permeation.* Materials Science of Membranes for Gas and Vapor Separationp. 159–190.
86. Wijmans, J. G. & Baker, R. W. (1995). The Solution-Diffusion Model: A Review. *Journal of Membrane Science, 107(1)*, 1–21.

87. Way, J. D. & Roberts, D. L. (1992). *Hollow Fiber Inorganic Membranes for Gas Separations.* Separation science and technology, *27(1)*, 29–41.
88. Rao, M. B. & Sircar, S. (1996). Performance and Pore Characterization of Nanoporous Carbon Membranes for Gas Separation. *Journal of Membrane Science, 110(1)*, 109–118.
89. Merkel, T. C., et al. (2003). *Effect of Nanoparticles on Gas Sorption and Transport in Poly (1-trimethylsilyl-1-propyne).* Macromolecules, *36(18)*, 6844–6855.
90. Mulder, M. (1996). *Basic Principles of Membrane Technology Second Edition.* Kluwer Academic Pub. 564.
91. Wang, K., Suda, H. & Haraya, K. (2001). *Permeation Time Lag and the Concentration Dependence of the Diffusion Coefficient of CO_2 in a Carbon Molecular Sieve Membrane.* Industrial & Engineering Chemistry Research, *40(13)*, 2942–2946.
92. Webb, P. A. & Orr, C. (1997). *Analytical Methods in Fine Particle Technology, 55,* Micromeritics Norcross, GA. 301.
93. Pinnau, I., et al. (1997). Long-Term Permeation Properties of Poly (1-trimethylsilyl-1-propyne) Membranes in Hydrocarbon—Vapor Environment. *Journal of Polymer Science Part B: Polymer Physics, 35(10)*, 1483–1490.
94. Jean, Y. C. (1993). *Characterizing Free Volumes and Holes in Polymers by Positron Annihilation Spectroscopy.* Positron Spectroscopy of Solids, 1.
95. Hagiwara, K., et al. (2000). *Studies on the Free Volume and the Volume Expansion Behavior of Amorphous Polymers.* Radiation Physics and Chemistry, *58(5)*, 525–530.
96. Sugden, S. (1927). Molecular Volumes at Absolute Zero. Part II. Zero Volumes and Chemical Composition. *Journal of the Chemical Society (Resumed)*, 1786–1798.
97. Dlubek, G., et al. (2004). *Positron Annihilation: A Unique Method for Studying Polymers.* in *Macromolecular Symposia.* Wiley Online Library.
98. Golemme, G., et al. (2003). *NMR Study of Free Volume in Amorphous Perfluorinated Polymers: Comparison with other Methods.* Polymer, *44(17)*, 5039–5045.
99. Victor, J. G. & Torkelson, J. M. (1987). *On Measuring the Distribution of Local Free Volume in Glassy Polymers by Photochromic and Fluorescence Techniques.* Macromolecules, *20(9)*, 2241–2250.
100. Royal, J. S. & Torkelson, J. M. (1992). *Photochromic and Fluorescent Probe Studies in Glassy Polymer Matrices.* Macromolecules, *25(18)*, 4792–4796.
101. Yampolskii, Y. P., et al. (1999). *Study of High Permeability Polymers by Means of the Spin Probe Technique.* Polymer, *40(7)*, 1745–1752.
102. Kobayashi, Y., et al. (1994). *Evaluation of Polymer Free Volume by Positron Annihilation and Gas Diffusivity Measurements.* Polymer, *35(5)*, 925–928.
103. Huxtable, S. T., et al. (2003). *Interfacial Heat Flow in Carbon Nanotube Suspensions.* Nature materials, *2(11)*, 731–734.
104. Allen, M. P. & Tildesley, D. J. (1989). *Computer Simulation of Liquids* Oxford University Press.
105. Frenkel, D., Smit, B. & Ratner, M. A. (1997). *Understanding Molecular Simulation: from Algorithms to Applications.* Physics Today, 50, 66.
106. Rapaport, D. C. (2004). *The art of Molecular Dynamics Simulation.* Cambridge University Press. 549.
107. Leach, A. R. & Schomburg, D. (1996). *Molecular Modeling: Principles and Applications.* Longman London.
108. Martyna, G. J., et al. (1996). *Explicit Reversible Integrators for Extended Systems Dynamics.* Molecular Physics, *87(5)*, 1117–1157.
109. Tuckerman, M., Berne, B. J. & Martyna, G. J. (1992). Reversible Multiple Time Scale Molecular Dynamics. *The Journal of Chemical Physics, 97(3)*, 1990.

110. Harmandaris, V. A., et al. (2003). *Crossover from the Rouse to the Entangled Polymer Melt Regime: Signals from Long, Detailed Atomistic Molecular Dynamics Simulations, Supported by Rheological Experiments.* Macromolecules, *36(4)*, 1376–1387.

111. Firouzi, M., Tsotsis, T. T. & Sahimi, M. (2003). Nonequilibrium Molecular Dynamics Simulations of Transport and Separation of Supercritical Fluid Mixtures in Nanoporous Membranes. I. Results for a Single Carbon Nanopore. *The Journal of Chemical Physics,* 119, 6810.

112. Shroll, R. M. & Smith, D. E. (1999). Molecular Dynamics Simulations in the Grand Canonical Ensemble: Application to Clay Mineral Swelling. *The Journal of Chemical Physics,* 111, 9025.

113. Firouzi, M., et al. (2004). Molecular Dynamics Simulations of Transport and Separation of Carbon Dioxide–Alkane Mixtures in Carbon Nanopores. *The Journal of Chemical Physics, 120,* 8172.

114. Heffelfinger, G. S. & van Swol, F. (1994). Diffusion in Lenard☐Jones Fluids Using Dual Control Volume Grand Canonical Molecular Dynamics Simulation (DCV☐GCMD). *The Journal of Chemical Physics,* 100, 7548.

115. Pant, P. K. & Boyd, R. H. (1992). *Simulation of Diffusion of Small-Molecule Penetrants in Polymers.* Macromolecules, *25(1),* 494–495.

116. Allen, M. P. & Tildesley, D. J. (1989). *Computer Simulation of Liquids* Oxford university press. 385.

117. Cummings, P. T. & Evans, D. J. (1992). *Nonequilibrium Molecular Dynamics Approaches to Transport Properties and Non-Newtonian Fluid Rheology.* Industrial & Engineering Chemistry Research, *31(5),* 1237–1252.

118. MacElroy, J. (1994). Nonequilibrium Molecular Dynamics Simulation of Diffusion and Flow in Thin Microporous Membranes. *The Journal of Chemical Physics,* 101, 5274.

119. Furukawa, S. & Nitta, T. (2000). Non-Equilibrium Molecular Dynamics Simulation Studies on Gas Permeation Across Carbon Membranes with Different Pore Shape Composed of Micro-Graphite Crystallites. *Journal of Membrane Science, 178(1),* 107–119.

120. Düren, T., Keil, F. J. & Seaton, N. A. (2002). *Composition Dependent Transport Diffusion Coefficients of CH_4/CF_4 Mixtures in Carbon Nanotubes by Non-Equilibrium Molecular Dynamics Simulations.* Chemical Engineering Science, *57(8),* 1343–1354.

121. Fried, J. R. (2006). *Molecular Simulation of Gas and Vapor Transport in Highly Permeable Polymers.* Materials Science of Membranes for Gas and Vapor Separation. 95–136.

122. El Sheikh, A., Ajeeli, A. & E. Abu-Taieh. (2007). *Simulation and Modeling: Current Technologies and Applications.* IGI publishing.

123. McDonald, I. (2002). *NpT-Ensemble Monte Carlo Calculations for Binary Liquid Mixtures.* Molecular Physics, *100(1),* 95–105.

124. Vacatello, M., et al. (1980). A Computer Model of Molecular Arrangement in a n-Paraffinic Liquid. *The Journal of Chemical Physics, 73(1),* 548–552.

125. Furukawa, S.-i, & Nitta, T. (2000). Non-Equilibrium Molecular Dynamics Simulation Studies on Gas Permeation Across Carbon Membranes with Different Pore Shape Composed of Micro-Graphite Crystallites. *Journal of Membrane Science, 178(1),* 107–119.

CHAPTER 12

INTERFACIAL TENSION OF WATER-IN-WATER BIOPOLYMER EMULSION CLOSE TO THE CRITICAL POINT

A. ANTONOV, P. VAN PUYVELDE, and P. MOLDENAERS

CONTENTS

ABSTRACT

Proteins and polysaccharides, being the main constructional materials in many biological structures, have a limited compatibility in aqueous media. At sufficiently high concentrations, they form water-in-water emulsions. Interfacial tension is an important parameter in such systems since it is a controlling factor in the morphology development during processing. In this work a rheo-optical methodology, based on the analysis of small angle light scattering (SALS) patterns during fibril break-up, is used to study the interfacial tension of water–sodium caseinate–sodium alginate systems located close to and relatively far from the binodal. The interfacial tension close to the critical point was $\sim 10^{-8}$ N/m, and it increased considerably, to a value of up to 5.2×10^{-6} N/m farther from the critical point. For the scaling of the interfacial tension with the density difference between the phases, a scaling exponent of 3.1 ± 0.3 was found, in agreement with the critical mean-field scaling exponent of 3.

12.1   INTRODUCTION

Biopolymer mixtures are of importance to control the composition–structure–property relationship in natural biological systems and in formulated foods [1–4]. The enormous diversity of food and other biological structures is based on only two kinds of macromolecules: proteins and polysaccharides, which are the most important constructional materials [5]. Water is a good solvent for each biopolymer, and it is known that their mixtures can display aggregative or segregative phase behavior. In the latter situation, each phase becomes enriched in either polysaccharides or proteins. Liquid–liquid phase separation in biopolymer mixtures is the rule, but it occurs at relatively high concentrations and at particular physicochemical conditions depending on the polyelectrolytic properties and the self-association behavior of the mixed macromolecules [6, 7]. The term 'water-in-water emulsion' was introduced to distinguish them from oil-in-water and water-in-oil emulsions [5]. These water-in-water emulsions are characterized by a very low interfacial tension [1, 8–10]. During the last 10 years, a significant amount of experimental (e.g., [11, 12] and theoretical studies (e.g., [13, 14]) have been performed to clarify the conditions of the incompatibility, to establish the origin of the interactions that lead to phase separation and even to predict phase diagrams. The main results of these studies were summarized in a number of reviews (e.g., [1, 15, 16]). If phase separation takes place, the driving force for decomposition must overcome the accompanying increase in interfacial free energy, which equals the product of the interfacial tension and the total interfacial area associated with the phase separation. Therefore the interfacial tension between the coexisting phases is of fundamental interest. In general, interfacial tension is of scientific and practical importance due to its relation to the thermodynamic state of biopolymer interfaces, its role in phase separation processes [17] as well as due to its relation to the morphology development in liquid two-phase fluid–fluid mixtures [17] The interfacial tension in aqueous biopolymer protein–polysaccharide mixtures is, in contrast to the phase equilibrium, an aspect to which limited attention has yet been paid. The magnitude of the interfacial tension depends on the position of the system on the phase diagram, tending to zero as the critical point is approached. Stokes et al. [19] applied shear to a two-phase gelatin–maltodextrin system, and from the shape of the droplets, they estimated an interfacial tension to be in the range of $(50–100) \times 10^{-6}$ N/m. Recently Guido et al. [20] developed an experimental methodology based on the analysis of the shape of a single droplet and reported very low interfacial tensions, of the order 10^{-5} to 10^{-6} N/m for sodium alginate–sodium caseinate systems. Van Puyvelde et al. [9] applied a rheo-optical methodology, based on small angle light scattering (SALS), to measure the interfacial tension of an aqueous gelatin dextran system located far from the critical point. They obtained a value of the order of 10^{-5} N/m. Recently, Scholten et al. evaluated the interfacial tension of gelatin–dextran–1 M NaI systems located far from and close to the critical point by the spinning drop method [10]. The value obtained was less than 10^{-6} N/m and it increased, in a nonlinear manner, farther away from the critical point. It should be noted that a highly concentrated NaI solution was used as a solvent to suppress the gelation of gelatin and to achieve small density differences between the coexisting phases.

The aim of this work is to determine the values of the interfacial tension in a two-phasic water–protein–polysaccharide mixture close to its critical condition, i.e. at phase compositions as close as possible to the critical point. The distance with respect to the critical point is expressed as the density difference between the coexisting phases, since at the critical point, this density difference becomes zero. A rheo-optical methodology [21, 22] based on the analysis of the SALS patterns during fibril break-up was used.

This method already proved to be well suited for measuring the interfacial tension in aqueous biopolymer mixtures [9] Here it will be explored how far the method can be extended to obtain values of the interfacial tension as close as possible to the critical point. In addition, a scaling relation for the interfacial tension will be presented that allows an estimate of the interfacial tension once the densities of the coexisting phases are known.

12.2   MATERIALS AND METHODS

The experiments are performed on a water–sodium caseinate–sodium alginate (w–c–a) system. These materials have been chosen because of their relatively large optical density which provides a large contrast for the light scattering experiment. Alginate is an anionic polysaccharide consisting of linear chains of (1–4)-linked D-mannuronic and L-guluronic acid residues. These residues are arranged in blocks of mannuronic or guluronic acid residues linked by blocks in which the sequence of the two acid residues is predominantly alternating [23] Casein is a protein composed of a heterogeneous group of phosphoproteins organized in micelles. Sodium caseinate is a salt of casein, obtained by disrupting the supra-molecular organization. This means that the polymer is a random coil and that the internal structure of the micelles is lost. The isoelectric point is around pH = 4.7–5.2 [24]. The caseinate at neutral pH is thus negatively charged, like alginate. Both biopolymers are well known, widely used in industry, and the thermodynamic behavior of the ternary water–caseinate–alginate systems is known from literature [6,25–28]. The sodium caseinate sample (90% protein, 5.5% water content, 3.8% ash, 0.02% calcium) was purchased from Sigma Chemical Co. The weight average molecular mass of the sodium caseinate is 320 kDa. The medium viscosity sodium alginate, extracted from brown seaweed (*Macrocystis pirifera*), was purchased from Sigma. The molecular weight of the sample, as reported in literature, was 390 kDa [20] Solutions of sodium caseinate and alginate were prepared by dispersing biopolymers in bi-distilled water under stirring and by heating for 1 h at 318 K. The solutions were then cooled to 298K and stirred again for 1 h. The required pH values of solutions (7.0) were adjusted by addition of 0.1–0.5 M NaOH or HCl. The resulting solutions were centrifuged at 60,000 × g for 1 h at 298K to remove insoluble particles. Concentrations of the solutions are determined by drying at 373K up to constant weight. The ternary water–caseinate–alginate systems with required compositions were prepared by mixing solutions of each biopolymer at 298 K. After mixing for 1 h, the systems were centrifuged at 60,000 × g for 1 h at 298K to separate the phases using a temperature-controlled rotor. The phase diagram of the ternary systems was obtained at pH 7.0, and 298K as illustrated schematically in Figs. 12.1a and 12.1b. The procedure is adapted from Koningsveld and Staverman [29] and Polyakov et al. [30]. The separately prepared protein and polysaccharide solutions (concentrations CO_2 and $C°_3$, respectively) were mixed at 298K in various weight ratios, yielding mixtures with protein and polysaccharide concentrations C_2 and C_3 (wt.%), respectively, which are represented on the phase diagram by the secant $[CO_2, C°_3]$. The phase state of the systems was determined by eye after centrifugation ($60{,}000 \times g$ for 1 h at 298 K) following a rest period of 1 h at 298 K. When phase separation occurred, the volumes of the two coexisting liquid phases were determined by weighing, assuming the density of the phases to be equal to 1.0. The points where the bimodal intersects the secant $[C°_2, C°_3]$ are given by extrapolation of the relation: $r = V¢/(V¢ + V¢¢) = f(C_2/CO_2)$ to $r = 0$ and 1. $V¢$ and $V¢¢$ represent volumes of the protein-enriched (lower phase) and the polysaccharide-enriched (upper phase) phases, respectively. The value $r = 0.5$ gives the position of the middle of the tie lines. By repeating this procedure for a series of values of CO_2 and $C°_3$, the phase diagram can be reconstructed from the set of points corresponding to the binodal and the center of the tie lines. The phase composition of the separated systems was derived from both the material balance of the systems studied and the position of the binodal in the plot, using the correlation [31]

$$\frac{V^I}{V^I} = \frac{\rho'X}{\rho'Y} \tag{1}$$

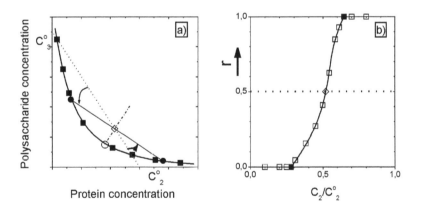

FIGURE 12.1 Schematic illustration of the phase volume ratio method. (a) Fragment of the phase diagram: dash-dotted line, secant; thick full line, binodal; thin full line, tie line; ●, phase composition; ■, points of the binodal; ○, critical point; dotted line, rectilinear diameter. (b) Typical dependence of the phase volume ratio on mixture composition.

where $V¢¢$, $V¢$ and $\rho¢¢$, $\rho¢$ are the volumes and density of the polysaccharide enriched and casein-enriched phases, respectively. X and Y are the distances in the plot between the points corresponding to (i) the composition of protein-enriched phase and the composition of initial mixture (X), and

(ii) the composition of polysaccharide-enriched phase and the composition of the initial mixture (Y). The phase composition of the separated systems was checked on the basis of the system's material balance. The threshold point was determined from the plot as the point where the line with the slope −1 is tangent to the binodal. The critical point of the system was defined as the point where the binodal intersects the rectilinear diameter, which is the line joining the center of the tie lines (Fig. 12.1). The resulting isothermal phase diagram of the w–c–a system is presented in Fig. 12.2, both in the Cartesian as well as in the classical triangular coordinates. The phase diagram was constructed to determine the critical point and is required to study the density difference and concentration dependence of the interfacial tension. Due to the high content of water in the coexisting phases and the strong difference in solubility, the phase diagram plotted in the Cartesian coordinates is more informative then the classical triangular representation.

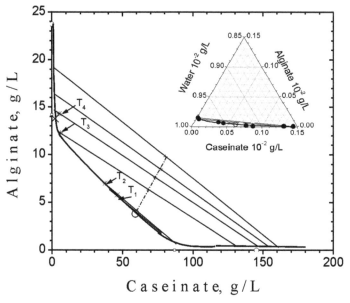

FIGURE 12.2   Isothermal phase diagram of the water–sodium caseinate–sodium alginate system (pH 7.0, 298 K). Coordinates of the critical point: caseinate, 5.91%; alginate, 0.38%.

The phase diagram is characterized by low values of the total threshold concentration (C^*_t = 16 g/L), a relatively high total concentration of biopolymers at the critical point ($C_{c,t}$ = 62.9 g/L) and a strong asymmetry (Ks = 15.5).

The viscosities of the alginate-enriched phases have been measured at 298K on a stress-controlled instrument (DSR, Rheometric Scientific) equipped with a Couette geometry. The viscosities of the caseinate-enriched phases have been determined with a Ubelhode viscometer at the same temperature. Gay–Lussac pyknometers have been used to measure the density of the coexisting phases. Each of the coexisting phases was measured at least four times to determine the statistical error. Flow small-angle light scattering experiments (SALS) measurements have been performed on a rheometrics optical analyzer (ROA) that has been modified to perform SALS measurements [32] A parallel plate geometry has been used and the temperature was controlled by means of a water bath at 298 K. For these experiments, the scattered light is intercepted on a screen that consists of a semitransparent paper with a beam stop. The resulting image is recorded by a CCD camera (Ikegami ICD-810P), which is mounted under the screen. The CCD camera is connected either to a frame grabber (Data Translation DT3851) or to a video recorder to collect the scattering patterns. Home built software was used to obtain intensity profiles and contour plots of the images (SALSSOFTWARE-K.U.L.).

12.3   RESULTS AND DISCUSSION

12.3.1.   RHEO-OPTICAL METHODOLOGY TO DETERMINE INTERFACIAL TENSION

The determination of the interfacial tension is performed by means of a recently developed rheo-optical technique [22] based on Tomotika's theory of fibril break-up [33] When a long fluid filament is present in a quiescent fluid matrix, interfacial instabilities due to thermal fluctuations will occur. These so-called Rayleigh instabilities will start to grow and will eventually disintegrate the fibril. Tomotika derived the following formula for the break-up time tB of a Newtonian fibril immersed in a quiescent Newtonian matrix [33]

$$t_B = \frac{2\eta_m R_0}{\Gamma\Omega} \ln \frac{\alpha_B}{\alpha_0} \qquad (2)$$

where α_B is the amplitude of the instability at break-up, α_0 is the amplitude of the initial disturbance, η_m is the viscosity of the matrix material (in this case the alginate-enriched phase), Γ the interfacial tension, R_0 is the initial radius of the undisturbed fibril that is breaking and Ω is a known function of the viscosity ratio of the mixture. Because of the dilute nature of the systems under investigation, the thermal value for α_B [34] will be used in the calculations leading to following expression for the break-up time:

$$t_B = \frac{\eta_m R_0}{\Gamma \Omega} \ln \frac{10^{23} \Gamma R_0^2}{T} \tag{3}$$

If all parameters are known, the interfacial tension can be calculated from Eq. (3) once t_B and R_0 are measured. The experimental flow protocol to generate fibrils consists of two parts and is schematically shown in Fig. 12.3. First the mixture is presheared to ensure a reproducible and well-defined initial morphology. The sample is allowed to relax for 10 s leaving enough time for full retraction of the deformed droplets. To create elongated fibrils a high shear rate is applied. This shear rate is chosen such that the resulting capillary number (representing the ratio of the hydrodynamic stress over the interfacial stress) largely exceeds the critical capillary number.

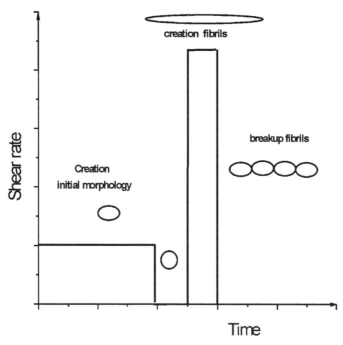

FIGURE 12.3   Schematic flow protocol used to create the elongated fibrils.

However, rather than going through the complete process of deformation, breakup and coalescence of the droplets, the flow can be stopped during the deformation process. Moreover, conditions can be created that ensure breakup by Rayleigh instabilities [35, 36]. During this disintegration, the evolution of the SALS patterns is monitored. It is known that during fibril breakup a peculiar scattering pattern emerges. It consists of two parts: central streak perpendicular to the prior flow direction is observed accompanied by secondary streaks. The intensity of these secondary streaks is increasing in time during the break-up process of the fibrils. Mewis et al. [21] demonstrated that the evolution of these scattering patterns could be modeled on the basis of Fraunhofer diffraction theory. More interestingly, important morphological information can be deduced from the patterns. From the scattering model, a simple relation between the polar angle of the secondary streak (θ_m) and the dominant wavelength of Rayleigh break-up λ_R can be found:

$$\theta_m = \arcsin(\frac{\lambda_1}{n_m \lambda_R}) \qquad (4)$$

with n_m the refractive index of the medium and λ_1 the wavelength of the light. According to Tomotika's model the relationship between the dominant wavelength and the initial fibril radius R_0 is determined by the viscosity ratio through [33:]

$$X_m = (\frac{2\pi R_0}{\lambda_R}) \qquad (5)$$

with x_m a known function of the viscosity ratio of the blend. Hence, from the SALS patterns, R_0 can be obtained experimentally. In addition Van Puyvelde et al. [22] and Mewis et al. [21] demonstrated that the moment at which the intensity in the second streak starts to level off can be identified with the break-up time tB of the fibrils. The leveling off or break-up point can be determined by identifying the maximal slope in the intensity versus time curve. With t_B and R_0 known, Eq. (3) can be used to obtain the interfacial tension. Various experiments can be performed varying both the applied shear rate and shearing time leading to a set of tB and $R0$ that are fitted to Eq. (3) to obtain the interfacial tension. In order to study the effect of the location of the system in the phase diagram on the interfacial tension, systems located along four tie lines have been studied as indicated on Fig. 12.2. Two tie lines (T_1 and T_2) were located very close to the critical point and two were located relatively far from the critical point (T_3 and T_4). The phase composition, viscosity and density of the coexisting phases of the various w–c–a systems are given in Table 12.1.

TABLE 12.1   Phase Compositions, Viscosity and Density of Coexisting Phases of the Water-Casein-Alginate System

Tie line	Alginate enriched phase				Casein enriched phase			
	Composition, g/L		Viscosity, Pa S	Density, g/L	Composition, g/L		Viscosity, Pa S	Density, g/L
	Casein	Alginate			Casein	Alginate		
I	42.7	6.00	0.088	1009.0	77.00	1.60	0.0247	1013.6
II	24.2	8.90	0.120	1008.0	87.00	0.80	0.0407	1014.3
III	4.47	12.40	1.220	1007.5	131.00	0.50	0.4532	1030.3
IV	2.00	14.50	1.920	1007.0	144.30	0.50	1.340	1040.0
V	1.30	16.30	Not determined	Not determined	153.40	0.50	Not determined	Not determined
VI	1.20	19.10	Not determined	Not determined	159.80	0.50	Not determined	Not determined

To create a blend with a droplet matrix morphology, a prerequisite to form the elongated fibrils during the flow protocol, 1% emulsions of the caseinate-rich phase in the alginate-rich phase have been used. The low concentrations of the dispersed phase are chosen to provide sufficient transparency necessary to perform the optical measurements.

The arrows presented in Fig. 12.2 represent the emulsions studied in this work. Fig. 12.4 shows the evolution of the SALS patterns of the two-phase system located the closest to the critical point (tie line T1) after an interrupted start-up in shear (shear rate of 3 s^{-1} maintained for 3 s). The corresponding microscopy images during the same flow history are shown in Fig. 12.5. During the deformation stage, the SALS pattern becomes highly anisotropic. This scattering pattern corresponds to highly elongated fibrils oriented in the flow direction as is shown in Fig. 12.5B. After stopping the flow, a secondary streak appears in the SALS images, indicative of the occurrence of Rayleigh instabilities. It should be noticed that the time, necessary to observe the development of the secondary streak, is rather long compared with the time-scales usually encountered in synthetic polymer blends [22] It can be noticed that the intensity in the secondary maximum is increasing with time and finally levels off when the fibrils are transformed to a string of separated droplets. From the analysis of this intensity as a function of time, the breakup time of the fibrils can be determined. During the disintegration process, the wavelength of the disturbance remains constant but its amplitude increases until the fibril finally breaks into a series of smaller droplets (Fig. 12.5D). Figure 12.6 shows the intensity in the secondary streak as a function of time ($t = 0$ corresponds to the interruption of the flow). The analysis of the SALS patterns is performed by the method explained above and elsewhere [9, 21, 22] leading—for the experiments represented in Fig. 12.4 to a fibril radius of 0.9 _m and a breakup time of 46 s.

With the break-up times and the values of $R0$ obtained from the SALS experiments, Eq. (3) can be used to obtain a value of the interfacial tension for the systems located along the four tie lines. For the system located along T1, the interfacial tension amounts to $1.2 \times 10^{-8} \pm 0.2 \times 10^{-8}$ N/m. The values for the other systems studied are given in Table 12.2. The rheo-optical method is capable of measuring these extremely low values of the interfacial tension, a value which would challenge most of the more classical interfacial tension methods. These results extend the applicability of the technique since it was originally developed for blends of synthetic polymers where the interfacial tension is of the order of 10^{-3} N/m.

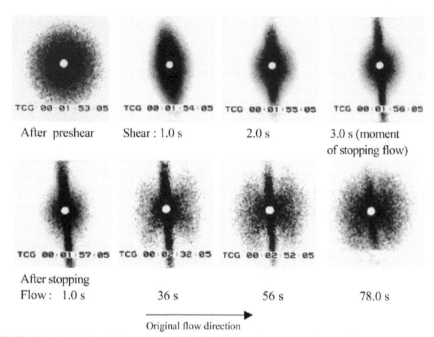

FIGURE 12.4   Evolution of the SALS patterns for an emulsion along T_1 after an interrupted start-up to a shear rate of 3 s^{-1}, interrupted after 3 s.

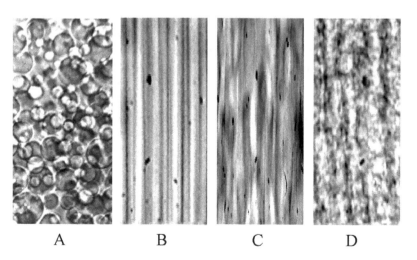

FIGURE 12.5   Evolution of the microscopy images, same conditions as in Fig. 12.4: (a) before application of the high shear rate; (b) moment of interrupting the flow (3 s at 3 s^{-1}); (c) 36 s after stopping the flow; (d) 76 s after stopping the flow.

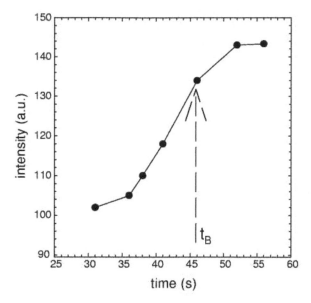

FIGURE 12.6   Intensity of the secondary streak as a function of time (same conditions as in Fig. 12.4).

TABLE 12.2   Interfacial Tension for the Systems Located on the Different Tie Lines

The line	Density difference (g/L)	Interfacial tension (N/m)
1	4.6	$1.2 \times 10^{-8} \pm 0.2 \times 10^{-6}$
2	6.3	$2.1 \times 10^{-8} \pm 0.2 \times 10^{-6}$
3	22.5	$1.2 \times 10^{-6} \pm 0.2 \times 10^{-6}$
4	33.0	$5.2 \times 10^{-8} \pm 0.2 \times 10^{-6}$

12.3.2   SCALING OF THE INTERFACIAL TENSION

In order to compare with theoretical predictions, the interfacial tension has been evaluated compared to the density difference between the coexisting phases. Obviously, at the critical point, the density difference $\Delta\rho$ between the coexisting phases equals 0. Hence the density difference is an indication of the distance of the system from the critical point. This approach is similar to the approach of Scholten et al. who used a spinning drop method to measure the interfacial tension of an aqueous gelatin/dextran mixture [10]. However, the values of the interfacial tension reported here is an order of magnitude smaller than the ones reported by Scholten et al. The distance from the critical point for the system studied by Scholten et al. is similar to the location of system T_1. To compare them on a relative scale, the density difference $\Delta\rho$ can be normalized by the density at the critical point ρ_{crit}. ρ_{crit} can be estimated, based on Table 12.1, as 1010 g/L, yielding a relative distance smaller than 0.005 which is similar to the values reported by Scholten et al. However, sodium iodide was added in their work to achieve a small density difference between the coexisting phases that was a prerequisite to make use of their spinning drop method, a drawback that is not present with this technique. In Fig. 12.7, the interfacial tension is plotted as a function of the density difference $(\Delta\rho)$ between the coexisting phases. The interfacial tension close to the critical point is very small $(1.2 \times 10^{-8}$ N/m) and increases to a value of $5.2 \times 10^{-6} \pm 0.15 \times 10^{-6}$ N/m farther from the critical point (T_4). The latter value for the mixture located on the phase diagram far from the critical point is in reasonable agreement to that obtained by Capron et al. $(8 \times 10^{-6}$ N/m) for a similar system [25] and similar to that reported by Guido et al. [20] $(8.8 \times 10^{-6}$ N/m) for a sodium caseinate (6 wt.%)–sodium alginate (1.0 wt.%) system located far from the binodal. As can be seen in Fig. 12.7, the dependence of Γ on $\Delta\rho$ is nonlinear. The data were fitted to the relation: $\Gamma = a(\Delta\rho)b$, for which $b = 3.1 \pm 0.3$ was found as best-fit parameter. The scaling exponent of 3.1 ± 0.3 is very close to that (3.0) predicted by mean-field theory [37]. A similar scaling exponent (2.7 ± 0.3) was obtained by Scholten et al. [10]for the gelatin-dextran–1 M NaI system. This scaling exponent is important since it allows to describe the interfacial tension in phase-separated biopolymer mixtures at different conditions provided the knowledge of one reference interfacial tension and the knowledge of the densities of the coexisting phases. This interfacial tension can then be used to describe for instance the phase separation or the morphology development during flow.

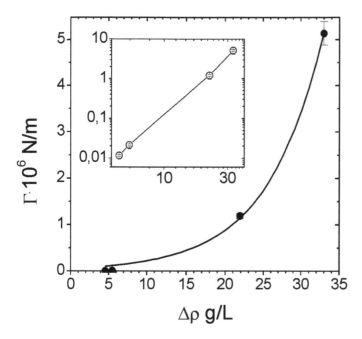

FIGURE 12.7   Interfacial tension of the systems along four tie lines as a function of the density difference. The full line is the best fit to the mean field scaling relation: $\Gamma = a(\Delta\rho)^{\delta}$ The insert shows the scaling plotted logarithmically.

12.4   CONCLUSIONS

Using a rheo-optical methodology, based on SALS measurement, the interfacial tension of a two-phasic water–sodium caseinate–sodium alginate system has been determined as a function of the location on the phase diagram. The interfacial tension close to the critical point was 1.2×10^{-8} N/m, and it increased to values of 5×10^{-6} N/m farther from the critical point. The rheo-optical methodology is very well suited to capture these very low values of the interfacial tension; the measurement window spans interfacial tensions ranging from 10^{-3} up to 10^{-8} N/m, which is hard to achieve with conventional techniques. The relation between interfacial tension and density difference between the phases has been determined and was compared with theoretical predictions. The experimental scaling exponent was in good agreement with critical mean-field values.

KEYWORDS

- **Biopolymer Emulsion**
- **Interfacial tension**
- **Rheo-optical methodology**
- **Small angle light scattering (SALS)**
- **Water-in-water emulsions**
- **Water–sodium caseinate–sodium alginate systems**

REFERENCES

1. Tolstoguzov, V. B. (1999) In: Roos, Y. H, Leslie, R. B, Liiford, P. J, Eds. *Water Management in the Design and Distribution of Quality Foods*. Lancaster, Basel: Technomic Publishing Company Inc.; 199–233.
2. Antonov, Y. A., Grinberg, V. Y., Zhuravskaya, N. A. & Tolstoguzov, V. B. (1980). Liquid 2-Phase Water-Protein-Polysaccharide Systems and their Processing into Structured Protein Products. *Journal of Texture Studies, 11*, 199–215.
3. Antonov, Y. A. & Soshinsky, A. A. (2000). Interactions and Compatibility of Ribuloso-1,5-Bisphosphate Carboxylase: Oxygenase from Alfalfa With Pectin in Aqueous Medium *Int. J. Biol. Macromol., 27*, 279–285.
4. Harding, S. E., Hill, S. E. & Mitchell, J. R. (1995). *Biopolymer Mixtures*. Nottingham: Nottingham University Press.
5. Tolstoguzov, V.B. (1986). In: Mitchel I.R, Ledward D. A, Eds. Functional Properties of Food Macromolecules. London: Elsevier Applied Science; p. 385.
6. Antonov, Y. A., Grinberg, V. Y. & Tolstoguzov, V. B. (1975). Phasengleichgewichte in Wasser. Eiweib. Polysaccharid-Systemen. Systeme Wasser. Casein. Saures Polysaccharid. *Starke, 27*, 424–431.
7. Antonov, Y.A., Grinberg, V.Y. & Tolstoguzov, *V.B.* (1976). *Polym. Sci. USSR* 1976, *18*, 566–569. [English translation].
8. Albertsson, P-A. (1985). Partitioning in Aqueous Two-Phase Systems: Theory, Methods, Uses and Applications to Biotechnology. Orlando: Academic Press.
9. Wolf, B., Scirocco, R., Frith, W. J. & Norton, I. T. (2000). Shear-Induced Anisotropic Microstructures in Phase-Separated Biopolymer Mixtures. *Food Hydrocolloids, 14*, 217–225.
10. Scholten, E, Tuinier, R, Tromp R. H. & Lekkerkerker H. N. W. (2002). Interfacial Tension of a Decomposed Biopolymer Mixturelangmuir, *18*, 2234–2238.
11. Antonov, Y. A. & Soshinsky, A.A. (2000). Interactions and Compatibility of Ribulose 1,5-Bisphosphate Carboxylase/Oxygenase from Alfalfa with Pectin in Aqueous Medium. *Int. J Biol. Macromol., 27*, 279–285.
12. Kasapis, S., Morris, E. R., Norton, I. T. & Gidley, M. J. (1993). Phase Equilibria and Gelation in Gelatin/Maltodextrin Systems. Part 2. Polymer Incompatibility in Solution. *Carbohydr Polym., 21*, 249–259.
13. Gottschaalk, M., Linse, P. & Piculell, L. (1998). Phase Stability of Polyelectrolyte Solutions as Predicted from Lattice Mean-Field Theory. *Macromolecule, 31*, 8407–8416.
14. Khokhlov, A. R. & Nyrkova, I. A. (1992). Compatibility Enhancement and Microdomain Structuring in Weakly Charged Polyelectrolyte Mixtures. *Macromolecules, 25*, 1493–1502.

15. Tolstoguzov, V. B. (2000). Composition and Phase Diagrams for Aqueous Systems Based on Proteins and Polysaccharides. *Int. Rev. Cytol.*, *192*, 3–31.
16. Antonov, Y. A. (2000). Concentration from Molecular-Dispersed and Colloidal Dispersed Solutions *Appl, Biochem, Microbiol.*, *36*, 325–337.
17. Funke, Z., Schwinger, C., Adhikari, R. & Kressler. (2001). Surface Tension in Polymer Blends of Isotactic Poly(propylene) and Atactic Polystyrene *J. Macromol Mater Eng*, 286,744–751.
18. Tucker, C. L. & Moldenaers, P. (2002). Microstructural Evolution in Polymer Blends. *Annual Review of Fluid Mechanics*, *34*,177–210.
19. Stokes, J. R, Wolf, B. & Frith, W.J. (2001). Phase-Separated Biopolymer Mixture Rheology. Prediction Using a Viscoelastic Emulsion Model. *J. Rheol.* 2001, *45*,173–189.
20. Guido, S., Simeone, M. & Alfani, A. (2002). Interfacial Tension Of Aqueous Mixtures Of Na-Caseinate And Na-Alginate By Drop Deformation In Shear Flow. *Carbohydrate Polym.*, *48*,143–148.
21. Mewis, J., Yang, H., Van Puyvelde, P., Moldenaers, P. & Walker, L. M. (1998). Small Angle Light Scattering Study of Droplet Breakup in Emusions and Polymer Blends. *Chemical Engineering Science*, *53*, 2231–2239.
22. Van Puyvelde, P., Yang, H., Mewis, J. & Moldenaers, P. (1998). Rheo-Optical Probing of Relaxational Phenomena in Immiscible Polymer Blends. *Journal of Colloid Interface Science*, *200*, 86–94.
23. Whistler, R. L. (1973). *Industrial Gums*, 2nd ed. New York: Academic Press; 1973.
24. Swaisgood, H. E. (1982). Chemistry of Milk Protein. Page 1. In P. J. Fox, *Developments in Dairy Chemistry 1*, 1–59, London: Appl. Sci. Publ.
25. Capron, I., Costeux, S. & Djabourov, M. (2001).Water in Water Emulsions: Phase Separation and Rheology of Biopolymer Solutions. *Rheol. Acta*, *40*, 441–456.
26. Antonov, Y.A., Grinberg V.Y. & Tolstoguzov, V.B. (1979). Thermodynamishe Aspekte Der Vertraglichkeit Von Eiweissen und Polysacchariden in Wassrigen Medien. Part.1. *Nahrung*, *23*, 207–214.
27. Antonov, Y. A., Grinberg, V.Y. & Tolstoguzov, V.B. (1979). Thermodynamishe Aspekte der Vertraglichkeit Von Eiweissen und Polysacchariden in Wassrigen Medien. Part 2, *Nahrung*, *23*, 597–610.
28. Antonov, Y. A., Grinberg, V. Y. & Tolstoguzov, V. B. (1979). Thermodynamishe Aspekte Der Vertraglichkeit Von Eiweissen Und Polysacchariden in Wassrigen Medien. Part 3. *Nahrung*, *23*, 847–862.
29. Koningsveld, R. & Staverman, A. J. (1968). Liquid-Liquid Phase Separation in Multicomponent Polymer Solutions. *Journal of Polymer Science*, *A-2*, 305–323.
30. Polyakov, V. I., Grinberg, V. Ya, & Tolstoguzov, V. B. (1980). Application of Phase-Volume Ratio Method for Determining the Phase Diagram of Water-Casein-Soybean Globulins System. *Polymer Bulletin*, *2*, 760–767.
31. Albertsson, P. Å. Partition of Cell Particles and Macromolecules, 3rd ed. New York: John Wiley & Sons; 1986.
32. Yang, H., Zhang, H., Moldenaers, P. & Mewis, J. (1998). Rheooptical Investigation of Immiscible Polymer Blends. *J. Polymer*, *39*, 5731–5737.
33. Tomotika, S. (1936). On the Instability of a Cylindrical Thread of a Viscous Liquid Surrounded by Another Viscous Fluid. *Proceedings of Royal Society of London*, *A150*, 322–337.
34. Kuhn, W. (1953). Spontane Aufteilung Von Fluessigkeitszylindern in Kleine Kugeln. *Kolloid Zeitung*, *132*, 94–99.

35. Stone, H. A. & Leal, L. G. (1989). Relaxation and Break-Up of an Initially Extended Drop in an Otherwise Quiescent Fluid. *Journal of Fluid Mechanics, 198,* 399–427.
36. Van Puyvelde, P., Moldenaers, P. & Mewis, J. (1999). Modeling and Scaling of Dichroism Relaxation in Immiscible Polymer Blends. *Physical Chemistry and Chemical Physics, 1,* 2505–2511.
37. Rowlinson, J. S. & Widon, B. (1984). *Molecular Theory of Capillarity.* Oxford, UK: Clarendon Press.

INDEX